Perspectives on Vietnam's Science, Technology, and Innovation Policies

Dao Thanh Truong

Perspectives on Vietnam's Science, Technology, and Innovation Policies

palgrave
macmillan

Dao Thanh Truong
VNU University of Social Sciences and Humanities
Vietnam National University, Hanoi
Hanoi, Vietnam

ISBN 978-981-15-0573-7   ISBN 978-981-15-0571-3   (eBook)
https://doi.org/10.1007/978-981-15-0571-3

This Palgrave Macmillan imprint is published by the registered company Springer Nature Singapore Pte Ltd.
The registered company address is: 152 Beach Road, #21-01/04 Gateway East, Singapore 189721, Singapore

For my wife, Dang Kim Khanh Ly, and my sons

PREFACE

Nearly 400 years ago, British philosopher, scientist, lawyer, and politician Francis Bacon (1561–1626) made the famous statement: "Knowledge is power." Unlike other resources that deplete when used, knowledge—especially pertaining to science and technology (S&T)—when shared and disseminated, has a multiplying impact. With great achievements resulting from the Industrial Revolution, Bacon's statement has never been more prescient. The role of S&T in human life today is undeniable. And just as predicted by Karl Marx, science has become the direct production force. With the development of scientific and technological revolutions parallel to industrial revolutions, the irreplaceable position of science is further affirmed. No country can develop without relying on S&T because they are the tools for nations to strengthen and enhance their positions in the global arena both economically and politically.

Innovation is the key factor determining a country's development. According to the global trend, businesses play a central role in the national innovation system. Science, technology, and innovation (STI) can help businesses make innovative products, processes, and services, but the government has to create an appropriate legal framework for this to take place. In many scholars' opinions, the current STI development policy of countries around the world is beyond the scope of traditional research and development (R&D) policy. Proper planning of STI policies is not a simple matter and must be considered under various aspects, especially how the country's STI system is operating. What, then, are the links among the factors in the system?

In recent years, Vietnam has made impressive achievements in economic and social development. Its gross domestic product (GDP) growth rate for 2018 was 7.08%, the highest increase since 2008. The contribution of total factor productivity (TFP) to GDP growth reached 43.5%, and the TFP average of three years (2016–2018) was 43.3%, much higher than the average of 33.6% from 2011 to 2015.[1] Vietnam's growing economy has helped to raise income, improve living standards, and create more opportunities for cooperation and development for the country. Vietnam officially became a member of the World Trade Organization (WTO) in 2007, joined the ASEAN (Association of Southeast Asian Nations) Economic Community (AEC) in 2015, and ratified the Comprehensive and Progressive Agreement for Trans-Pacific Partnership (CPTPP) in 2018. Ironically, these achievements also pose a challenge for the next development orientation of Vietnam as the economy strives towards high and sustainable growth. Vietnam is at the crossroads of development. On the one hand, it continues to exploit the potential in areas such as increasing investment capital, labor, natural resources, land, and so on. On the other hand, the country needs to develop and promote S&T, the legal corridor, and the quality of human resources.

The role of S&T development in Vietnam was soon confirmed in the country's legal documents. For example, Article 62 of the 2013 Constitution stipulates that "S&T development is the leading national policy, playing a key role in the country's socio-economic development".[2] In addition, the main task of the government's Action Program to implement the socio-economic development strategy for 2011 to 2020, as well as the five-year directions and tasks of national development from 2011 to 2016, articulates: "Science and technology development is really a key driver of fast and sustainable development."[3]

In recent years, interconnection and cooperation between the academic sector (research institutes and universities) and enterprises in Vietnam have been limited. S&T have not met current social needs and demands, and have not been effective tools to implement the national strategic tasks. A number of policy measures have been proposed and conducted to overcome these issues, and these have seen some success. However, the success is only reflected in some outstanding cases at a narrow scope. The new STI policies are still traditional in the sense that the goal of developing R&D activities is the main task, such as commercializing research results from S&T organizations and encouraging institutes/schools to associate with businesses. Proper attention has not been paid to business innovation activities. Despite being at the center of the national innovation system, enterprises' innovation is still not the focus.

So, what is necessary for Vietnam to rise?

It would require the planning and implementation of appropriate STI policies, which play a very important role for the development of STI, as well as the integration of international services for S&T development in particular and the national economy development in general.

In terms of research, there have been many research projects on S&T systems, R&D systems, innovation policies, and national innovation systems in Vietnam. In addition, there are studies on various aspects of innovative policy, such as excellence centers, science parks, S&T enterprises, innovative cluster, and innovation management. However, the overall picture of the STI system and the system development policies has not yet been reviewed and analyzed deeply. Hence, serious research on the situation and proposing solutions to develop STI systems in Vietnam is necessary.

This book is based on the research results of the two projects: "Research and Analysis of Vietnam's Science, Technology and Innovation System in the Trend of International Science and Technology Integration" (Code KX06.06/11-15)[4] and "Management Policies on Social Mobility of High-Quality Science and Technology Human Resources of Vietnam in the Context of International Integration" (Code KX01.01/16-20),[5] carried out by Assoc. Prof. Dao Thanh Truong as the project manager. Some data in this book have been published (in Vietnamese) by Thế Giới Publishers in 2016, however the author has edited and upgraded it to provide a thorough version. The book is a collection of the author's desires, concerns, and hopes for Vietnam's S&T development in general and Vietnam's STI policy in particular. The author outlines a general picture of the current status of the STI system in Vietnam, thereby frankly acknowledges the shortcomings, and proposes policy solutions to each type of S&T institution. The author hopes that the insights can advance Vietnam's STI development goals in the context of international integration.

The book examines the concept of STI, provides an overview of research works, and introduces the experiences of some countries regarding building and developing STI systems. Significantly, the book shows the reality of the *force* and *quantity* of each component of the S&T system (research institute, university, and enterprise), using the analysis and evaluation of survey data. This work aims to analyze the current status of STI policies and identify the philosophy of existing policies to provide recommendations and orientations on macro-level policies.

Errors are inevitable in any book. The author looks forward to receiving comments from managers, researchers, interested people, and readers everywhere. The support and exchange of ideas will be the motivation for the author to conduct further studies on Vietnam's STI system in the international integration process.

Hanoi, Vietnam   Dao Thanh Truong

Notes

1. General Statistics Office (2018), Report on socio-economic situation in 2018. Retrieved from: https://www.gso.gov.vn/default.aspx?tabid=621&ItemID=19037, accessed on 27 December 2018.
2. National Assembly of the Socialist Republic of Vietnam (2013), The Constitution of the Socialist Republic of Vietnam (amended). Article 62.
3. Government of the Socialist Republic of Vietnam (2012), Resolution promulgating the Government's Action Program to implement the socio-economic development strategy for 2011–2020 and the directions and tasks of national development in the five years from 2011–2015, issued on 24 April 2012.
4. This national-level project was implemented from 2013 to 2015 in the framework of the national key S&T program KX06/11-15 on the international integration of S&T.
5. This national-level project was implemented from 2016 to 2019 in the framework of the national key S&T program KX.01/16-20 on key issues in social sciences and humanities for Vietnam's socioeconomic development.

Contents

LIST OF FIGURES

LIST OF TABLES

LIST OF BOXES

Some Theoretical Issues About STI and STI Research in the Trend of International S&T Integration

1.1 Overview of Research on STI in International S&T Integration Tendency

1.1.1 Foreign Research on STI System

Research on innovation was pioneered in 1903 by Gabriel Tarde, who first drew the S-shaped diffusion curve diagram. Tarde determined the innovation decision with the following steps: initial knowledge, forming an attitude, decision to approve or disapprove, implementation and use, and confirmation of the decision.

From the early years of the twentieth century, economist Joseph Schumpeter had conducted research on innovation by distinguishing the boundaries between the formation of product ideas or processes and the use of such ideas in economic development, which was published in his book in 1911. Initially, the STI system was mentioned in studies of innovation and the national innovation system (NIS) by Freeman et al. (1982), Lundvall (1992), Nelson and Rosenberg (1993), Carlsson and Stankiewicz (1991), and Edquist (1997). Innovation was understood as a broad concept that covered processes employed by businesses and included the design of new products and processes of construction for businesses, regardless of whether they were new locally or internationally. The concept of innovation was not only about new technology, but also the spreading of that technology. This meant that innovation was considered primarily as a result of interactive

© The Author(s) 2019
D. T. Truong, *Perspectives on Vietnam's Science, Technology, and Innovation Policies*,
https://doi.org/10.1007/978-981-15-0571-3_1

learning processes. Knowledge interacted in new ways to create new knowledge, processes, and products. Such interactions not only pertained to R&D, but also to production and business activities. Interactions occurred within a business, between businesses and consumers, among businesses, or among businesses and public organizations.

Scholars have focused on theoretical research and methods of forming innovative systems at different levels: countries (macro), regional (mezo), and businesses (micro), as well as the relationships between these levels. Studies show that the innovation system is the best way to overcome the shortcomings of the developed market model under the free-market mechanism. The national and regional innovation system is a systematic linkage of institutions used to create competitive new products in the market, where enterprises play a central role and the government plays a supportive role (by legal corridor). This is a prerequisite for forming an innovation system. One commonality among these studies on innovation systems is that innovation is recognized as a multifaceted phenomenon that cannot be easily included in a particular industry. Fagerberg et al. (2005) pointed out important elements of that system, including businesses, innovators, innovation process, organization, and innovation assessment. Researchers also argue that innovation is seldom held in an independent manner, but interacts with other aspects—such as R&D, policy, education, finance, economic growth, employment, intellectual property, and the differences of regions, industries and multinational organizations—that affect innovation. In addition, the study acknowledges technological quality as the basis for the penetration of innovation into all economies. It also concludes that low- and medium-tech industries are less innovative than high-tech industries.

NIS studies by Charles Edquist (2001) presented different approaches to the innovation system: economic differences between countries account for differences between innovation systems and showed the way for promoting faster and more effective technological innovation. Franco Malerba (2002) posited that the major factors affecting innovation are knowledge and technology, agents and networks, and institutions.

J. Hauknes and O. Wicken (STEP Group) (1999a) reviewed some basic characteristics of innovation policy, stages of innovation policy, and regulatory tools for industrial development. Studies by Edquist (Linköping University, Sweden) in 1999 and 2001 mentioned innovation policy as well as the design and implementation of innovation policies within the framework of an innovative system approach. The study conducted by the

Charles River Association (Asia Pacific) (2003) looked into Singapore's innovation policy in terms of foreign investment, venture capital, R&D policy, cluster policy, and links. The Swedish Institute for Growth Policy Studies (ITPS) (2004), in collaboration with the Swedish embassy in Tokyo, focused on current issues in Japanese research and innovation policies as well as government funding trends for R&D.

In addition to programs associated with the national technology innovation, a number of regional technology research and innovation programs are also offered in some countries. For instance, the Belgian RIT program aims to encourage Wallonia small and medium-size enterprises (SMEs) to conduct R&D activities and technological innovation through recruiting special innovation managers. Hungary's Baross Gábor program supports technological innovation in the Dél-Alföld region to enhance knowledge transfer between R&D institutes and businesses to promote the commercialization of research results. These are done through support measures that cover developing products and technologies for small businesses, facilitate cooperation between businesses and other organizations in innovation activities and knowledge exchange, and strengthen the commercialization of results of innovation activities in enterprises.

Bengt-Ake Lundvall et al. (2006) presented four very specific, profound case studies on the innovation system of Asia, specifically those of Thailand, Hong Kong, Indonesia, and Korea. These four Asian countries have experienced and are still facing some common and major issues. This shows that the government plays a very important role in establishing prerequisites for the transition. It is referred to as a "gardener" that supports innovators by providing proper financing and other measures, as well as removing cumbersome paper procedures, institutional and competition barriers to innovation, and increasing investment-based knowledge in education and research. The study also indicates the opportunities and challenges caused by social, economic, and political division as well as ethnic issues.

In the innovation system, enterprises are considered to be central actors in the innovation process. They are responsible for bringing in new things, as well as implementing and creating values. In competitive theory, Michael Porter (1990) acknowledged the competitive advantage established through innovation. The "innovative machine" is needed to produce new things that help businesses gain a competitive advantage to step into the future.

The issue of what enterprises must do to execute the innovation process was acknowledged very early on and mentioned in Richard Foster's 1986 work, *Innovation: The Attacker's Advantage*. He referred to innovation as an advantage for those who actively create change. This view was shared by Fagerberg et al. (2005). When exploring the nature of innovation, they considered enterprises as important factors that are located within relationships and depend on other organizations such as businesses (suppliers, customers, and competitors) and non-business organizations (universities and government departments).

As businesses are growing with a variety of goods and services, innovation plays the core role in the overall business strategy. Identifying innovation strategies is an effective way to help firms win in a competitive environment. The innovation strategy focuses on solving three main questions: What? (What products and services can satisfy customers?), Who? (Which customers want to buy your products and services?), and How? (How do businesses satisfy customers?) (Möller et al. 2008). Businesses need to advance their development by anticipating demand and creating demand for customers (Palmer and Kaplan 2007). This will lead to the formation of a new market with several prospects and challenges for businesses. Regarding the development of enterprises, innovation allows enterprises to gain their own position in the market. It brings value and opens up avenues for businesses to build facilities to compete and win. Innovation is what every business needs for its development (Hammer 2006).

It is clear that globalization affects all sectors and activities, including innovation in countries. Several issues are raised—such as the relationship between globalization and innovation; how globalization affects the innovation system in different countries both positively and negatively; and how innovative policies are deployed in the global context with complexity, dynamism, and unlimitedness. Richard Nelson (1993) recognized that the innovation system is not limited by national boundaries. The commercialization of high-tech products by companies in industries—such as semiconductors, computers, telecommunications, aviation, and biotechnology—special government support policies and the cross-nationalization of industries have become matters of great concern among nations. Fred Gault (2010) mentioned the issues of access to innovation in another aspect: how to measure innovation, and how policies have been planned and implemented to support the innovation process. The development of an indicator system to measure innovation activities in the economy, which provides the numbers and indicators used to monitor and assess the impact of policies, has become an urgent issue in many nations. Because of these activities, innovation and innovative policies are better understood, thus also bringing better results and impact on the economy and society. Gault also offered very valuable perspectives: R&D is not innovation unless it is connected to the market, so are invention and research publications. Furthermore, the author emphasizes intellectual property as part of the innovation policy. This book aims to propose actions to be taken in the short and medium terms for innovation strategies by focusing on high, multidisciplinary qualifications, accelerating innovation, and sustainable production growth to cope with global challenges.

1.1.2 Domestic Studies on STI System

Theoretical studies on innovation, innovative policies, and national innovation systems in Vietnam are conducted based on foreign sources as cited above. Some examples are Hoang Van Tuyen (2006) and Nguyen Thi Minh Nga (2006). While the state and the community have recently paid more attention to innovation, this idea is often associated with activities such as startups. The wave of research and interest in STI has become stronger. One can also mention the research carried out by international organizations such as the Organization for Economic Co-operation and Development (OECD) and the World Bank, or domestic organizations and researchers such as the Central Institute for Economic Management, General Statistics Office, Phung Xuan Nha and Le Quan (2013), Dang Ngoc Dinh (2015), and Pham Ngoc Minh (2015).

For a long time, researchers in Vietnam have avoided the phrase "national innovation system". Instead, other expressions would be used in official documents, such as "national innovation system for science and technology" and "national innovation system by science and technology". The main reason for this is probably the fear of confusion between the Doi Moi reforms of 1986 and the innovation in S&T. The remainder of this section explains the concept of innovation in more depth.

In Vietnam, components of the national innovation system and the innovation policy are already in place. Some policies for S&T were even seen as a milestone of innovation. However, that system has not been formed, and the policy for that system is not yet implemented. This is repeated several times in some S&T policy analysis studies. It is worth noting that a policy review report for the development of the S&T development strategy for the period 2000 to 2010 was developed by an expert team from the International Development Research Centre, Canada, through interviews with businesses and managers, scientists and some social activists in Vietnam's northern, central, and southern regions. The report has given some remarkable conclusions:

- Vietnam has too many published policies and implicit policies, and most of the implicit policies are found to hinder published policies.
- Vietnam has all the components of the national system, but this system has not been completely established.
- Many policies are overlapping and invisible, and this impedes the effectiveness of policy implementation. This means that innovation policy is not available.

The overlapping of S&T policies is mentioned by several policy researchers such as Nguyen Thi Minh Nga (2006), Nguyen Hong Ha (2004), and Nguyen Van Hoc (2005).

In some other research works, innovation is emphasized in relation to R&D activities and issues have emerged from these activities. For example, Tran Ngoc Ca (2006) found that, while Vietnam spent only about 0.5% of its GDP on R&D activities in 2003, most countries in the OECD and China spent about 2% of GDP for the same purpose. Moreover, most R&D grants in Vietnam are offered to government research institutes. Only a very limited number of universities have the resources needed for R&D activities.

Tran Ngoc Ca and Nguyen Vo Hung's (2012) study, which is included in the Vietnam–Sweden research cooperation program on innovation system, provides a realistic view on some industries in the agricultural sector, which Vietnam is strong in. The work shows a general picture of SMEs in Vietnam with a low level of technology, human resources, technical equipment, and infrastructure, as well as a low level of investment in technology innovation compared with development requirements, and very little R&D activities. In such a context, the innovation and technology transfer activities of SMEs are even worse. Due to limited capital, an undeveloped technology market, and weak scientific information, most SMEs' equipment in the field of processing agricultural products is produced domestically (tea accounts for 76%, vegetables and fruits 87%, and coffee 86%). These devices are often acquired from state-owned enterprises, patched and repaired. The majority of the equipment are old technologies imported from the Soviet Union, China, and India, while a small number comes from Japan and Taiwan.

According to Tran Ngoc Ca, R&D activities in the SME sector are generally weak. R&D resources are just a formality, relying less on R&D organizations and universities, and large businesses have a greater advantage to do so. Enterprises with foreign direct investments tend to rely on R&D organizations in their own countries. Therefore, there are very few opportunities for R&D organizations and universities to play an important role in helping these businesses. The relationship among universities, research centers, and businesses, and between businesses and universities in particular, still entails many barriers. This weakens the efforts of universities to better serve the needs of the local economy. As such, it can be concluded that there is a need for technology and training services to be provided by institutes and universities. However, these needs are difficult to meet, and the expansion of relationships among businesses and universities is not satisfactory. The contributions of R&D and training of universities are still below the desired level.

Domestic and foreign studies show the importance of STI activities and STI policies in socioeconomic development. STI has long been a criterion for evaluating the quality of growth of nations. Studies have focused on the role of innovation, indicators of innovation, comparison of innovation systems, and so on. However, the gap in STI policies—especially the assessment of STI policies and the impact of STI policy impact, as well as the development of STI policy in a context of such extensive integration—is still left open. Currently, STI development trends in the world are positively impacting Vietnam's development choices over the years and in the future. Vietnam is moving away from a centrally planned economy to a multisector market economy, according to social needs. This is considered a move to "rearrange the country"[1] after months in crisis and stagnation. From a poor, low-income country, Vietnam has gradually stepped out from the postwar ruins to build an image of a country with a high growth rate, belonging to a group of middle-income countries, and step by step integrating extensively with the region and the world. This is a continuous effort. Vietnam is in the process of integration and, of course, changes in STI management policy will be implemented under the inevitable trend of the market economy, in line with the global STI development trend. This move is a stepping stone for Vietnam to integrate more deeply with the world.

1.2   The Theory of STI System in the Context of International S&T Integration

1.2.1   Concept of Innovation

Some argue that Charles Edquist, a professor at Sweden's Linköping University, proposed the concept of "system of innovation". However, that is untrue, as it was not until the 1990s that Edquist began writing about innovation systems.[2]

Meanwhile, from the late 1970s, the Hungarian Technical Development Committee had formed a research team on innovation system called the Innovation Team, or Innteam. J. Vecsenyi presented the report, "An Innovation Policy Analysis in Hungary", at the Seminar on Innovation Management, held at the Institute of Analysis of International Application Systems in Laxenburg, Austria, from 22 to 25 June 1979.

Going back further in history, it can be seen that the concept of innovation was defined and its role clarified in the first economic development in Joseph Schumpeter's *The Theory of Economic Development: An Inquiry into Profits, Capital, Credit, Interest, and the Business Cycle* (1911, 1934 [trans.]). Researchers later cited his idea of innovation in the studies of economic sectors and in the fields of technological innovation in production.

In the 2014 survey of innovation studies by Edison et al. (2013), 40 definitions of innovation were introduced. Among them, the OECD (2005) defines an innovation to be the implementation of a new or significantly improved product (goods or service), a process, a new marketing method, or a new organisational method in business practices, workplace organization, or external relations.[3]

Based on the OECD's definition, Crossan and Apaydin (2009, p. 1155) define innovation as: "[the] production or adoption, assimilation, and exploitation of a value-added novelty in economic and social spheres; renewal and enlargement of products, services, and markets; development of new methods of production; and the establishment of new management systems. It is both a process and an outcome."[4]

The International Organization for Standardization (ISO) (2015) defines innovation as a "new or changed object, realizing or redistributing value".[5]

Section 16, Article 3 of the Vietnam Law on Science and Technology (2013) defines innovation as "the creation and application of achievements, technical solutions, technologies, management solutions to improve the efficiency of socio-economic development, improve productivity, quality and added value of products and goods".

According to the above definitions, innovation is easily misunderstood as invention. But these are two completely different concepts. Both innovation and invention relate to new things. However, invention refers to the creation of a new product, while innovation refers to the value of a new product.

1.2.2   STI System

The concept of STI system was introduced to Vietnam in the 1980s with presentations by Hungarian researchers. It was being researched from 1990 to 1995 through the Swedish Agency for Research Cooperation (SAREC) projects by Nguyen Thanh Ha et al. (1992–1995).[6]

While "STI system" stands for "science, technology, and innovation system", other studies call this "science, technology, and creative innovation system". Still others in Vietnam refer to innovation as "innovation/creation" to avoid confusion with the concept of Doi Moi, meaning "reform"—a term introduced in 1986 in the documents of the Communist Party of Vietnam. However, another remarkable thing is that the website Encyclopedia.com refers to Vietnam's Doi Moi process as "Renovation" in English.[7] In the Online Dictionary, we also see examples of the definition of "Renovation" (French) as in "Renovation in Vietnam".[8] "Doi Moi" in English documents in Vietnam is called "Renovation" or "Reform", rather than "Innovation". Encyclopedia.com writes: "Doi Moi (Renovation in English) is the name for reform initiated in Vietnam in 1986."[9]

Thus, semantically, Doi Moi should be called "Reform" or "Renovation". In Japan, the reform carried out by the Meiji Emperor in mid-nineteenth century was called "Restoration", as in "Meiji Restoration".[10]

According to B. Carlsson and R. Stankiewicz (1991), the technology innovation system is a concept developed in the scientific areas of study innovation to explain the nature and rate of technology change. A technological innovation system can be defined as "a dynamic network of interaction agents in a specific economy/industry sector under a specific institutional infrastructure and participation in stages, diffusion and use of technology" (Carlsson and Stankiewicz 1991, p. 93).[11]

Hekkert et al. (2007) define the development innovation system as an innovative system approach that focuses on explaining the nature and speed of technological innovation. The innovation system is thus said to be a set of components and rules that affect the speed and direction of technological change in a particular area of technology. A common definition of an innovation system is a group of components (devices, objects, or facilities) that serve a common purpose, that is, to work toward a general or objective function. The components of an innovation system are factors, networks, and organizations that contribute to the overall function of development, diffusion, and use of new products and processes (Bergek et al. 2008). A system can be a collective and coordinative action, so a system innovation is not necessarily a structural analysis that illustrates better motivation and system performance but also the interactions between systemic components.

In Vietnam, many seminars state that the concept of innovation system carries a general value, which should be used, instead of the concept of the S&T system, or the concept of an R&D system. However, such an understanding is not really appropriate. These concepts are not interchangeable at all. Each concept has its own role.

Under the system approach, the STI system is identified as follows (Dao Thanh Truong 2016):

1. The innovation system implies *the purpose* of the system. Such a purpose is innovation.
2. The S&T system is related to the *means of* innovation, meaning science and technology.
3. When the R&D system is related to the *core means* to innovate, it is "research and deployment".
4. When visualizing the generalization, the STI system is a system with full innovation purposes, using the means of S&T and the core means of research and deployment.

STI factors include not only businesses, universities, and research institutes, but also other organizations such as associations and non-commercial organizations. The STI system consists of a set of constituent elements that interact in pairs with M (Market)—R (Research)—D (Deployment) and—P (Production). Therefore, STI policy has a quite wide scope, covering not only S&T. Above all, STI policy not only fits within the S&T policy framework, but also other related policies such as tax, finance, business, intellectual property policies, etc. STI policy itself is both perfect (old/existing ones), adaptive (under environment) and self-renewal (under the requirements of itself, by the nature of STI).

1.2.3 Structure of the STI system

As defined by the OECD, STI is a system of public and private agencies whose activities aim to explore, import, transform, and disseminate new technologies. It is a reciprocal system of public and private enterprises, universities and government agencies, aimed at the development of S&T within the country. The reciprocity of these units can be technical, commercial, legal, and financial, for the purposes of developing, sponsoring, or implementing S&T activities.

The elements of STI include not only enterprises in the entire value chain (including large businesses and SMEs), universities and research institutes, but also public organizations, other related organizations such as industry associations and non-commercial organizations, venture capital organizations, standards organizations, etc.

According to the OECD, the STI system is modeled as shown in Fig. 1.1.

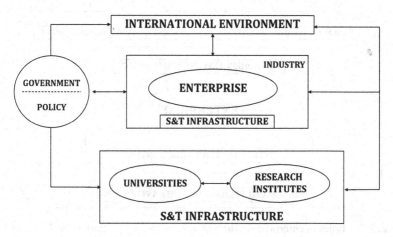

Fig. 1.1 STI model by OECD. (Source: Nguyen Van Hoc 2012, Lecture on National Innovation System, Faculty of Management Science. VNU University of Social Sciences and Humanities, Vietnam National University, Hanoi, Vietnam)

Based on the OECD model, there are two main generation components: *elements* and *the relationship between elements* in the STI system. Elements include:

- *Government*: The government has a macro management role, regulating activities in the STI system. The government monitors, inspects, and develops policies and plans relating to STI activities; allocates resources and budgets to S&T industries and activities according to priority; sets up incentive programs to promote innovation and other S&T activities; ensures the ability to implement policies and coordinate activities; guarantees the ability to forecast and evaluate trends of technological change; establishes, operates, and maintains information operation policies, general S&T equipment facilities; and sets up national measurement, standards, and intellectual property systems.
- *Enterprises*: Enterprises are considered to be agents of innovation. Enterprises quickly develop the market, ensure the specialization of the market, enhance the value of S&T products, make socio-economic strategies, and then return to strengthen investment in resources for STI activities. The role of enterprises is described in Fig. 1.2:

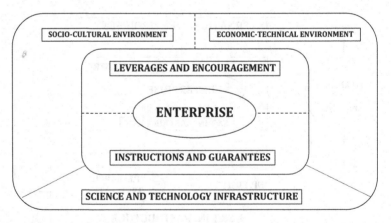

Fig. 1.2   Roles of enterprises in the STI system. (Source: Nguyen Van Hoc 2012, Lecture on National Innovation System, Faculty of Management Science. VNU University of Social Sciences and Humanities, Vietnam National University, Hanoi, Vietnam)

- *Universities*: In the university environment, the commercialization of R&D products often takes place in three strategic approaches, namely the sale of patent use rights, the implementation of scientific research contracts on order, and the establishment of university companies (Kroll and Liefner 2008).[12] The importance of research and deployment is shown by the creation of new technology companies (Hindle and Yencken 2004). Moreover, the research conducted at universities/research institutes has increasingly attracted the attention of industries. Today, industries are relying heavily on the research and technological development results of universities Ab Aziz et al. (2011). This area has a role in the orientation of training and research for S&T human resource for STI activities, creating a new knowledge source.
- *Research institutes*: They play a role in the development of basic research and applied research, contributing to S&T knowledge sources. This area will use S&T trained human resources and "demand" research topics from enterprises. At the same time, this area provides knowledge for the university sector and technology products for the business sector.
- *Environment*: In addition, innovation activities need a series of environmental conditions for innovation. These are the conditions of the information infrastructure, technology infrastructure, and industrial infrastructure of innovation.

Foremost, information is needed to innovate.

Here, *information* includes that relating to the world's trend of innovation, technological innovation, and innovation in management, business, and market. Information is the raw material for innovation. Information answers the question, "What is renewing?" The innovation activity would have no content without information.

Technology—more specifically, the technological capacity of the country as a whole, and of each facility in particular—is a basic condition to create a basis for technological innovations. There must be the technological capacity to acquire technology. Technological capacity contains skills of human resources, technological equipment capacity, and technological information capacity.

Finally, *industry* as described here is the industry of manufacturing technology equipment. In the absence of industry involvement, all technology principles would not be realized.

In terms of contacts, there are:

1.2.3.1 Cooperation Activities Between Enterprises

Since the enterprise is the area to implement R&D policies and innovation sources in OECD economies, one of the most significant links in STI is derived from technology cooperation among enterprises as well as relations to get more information.

1.2.3.2 Academic–Industrial Relations

Another relationship in the STI system is the link between academic sectors (research institutes and universities) and industrial parks (including enterprises and production supporters). The quality of research infrastructure and its links to industry are among the most important factors for supporting innovation. Linking activities also respect the regional or local elements. Knowledge flows from the research sector to the industry may be important to a particular locality, but may be unimportant for other regions or localities. There is a trend that recognizes the creation of "knowledge centers" near top universities that are oriented to research and developing special technologies such as biotechnology, computer software, and information technology. For example, the United States has Silicon Valley in California (near Stanford University and the University of California), Boston's biotech cluster (near the Massachusetts Institute of Technology) and telecommunications in New Jersey (near Princeton University and Bell Laboratories).

1.2.3.3 Technology Dissemination

Disseminating traditional technology is the dissemination of technology in the form of new equipment and machinery. However, the innovation of enterprises depends on the installation and operation from the acquisition technology, and the use of new products developed elsewhere. Knowledge of technologies can be provided by universities, research institutes, etc. The diffusion of technology is particularly important in the manufacturing field and service industries, which cannot carry out R&D themselves.

1.2.3.4 Mobilizing and Using S&T Human Resources

S&T human resources, especially high-quality S&T personnel, play an important role in controlling all activities in the STI system. Present in every element, S&T human resources are the management objects and are affected by STI policy. Formation and development of "underground knowledge" flow are also activities in S&T human resources, contributing to knowledge transfer among areas of the system.

As such, analyzing components and the relationship between elements of the STI system is important for policymakers to create a mechanism that reduces the gap among universities, research institutes, and businesses to flexibly circulate factors such as manpower and finance across these areas. This can encourage businesses to play a central role in innovation activities, invest in STI activities, and improve competitiveness in the context of increasingly deep S&T integration in a volatile world.

1.2.4   STI Policy

There have been many documents, especially by the European Commission, and by some Western scholars studying innovation that offer different perspectives on innovation policy. Stoneman (1987) considered innovation policy as that relating to government intervention in the economy with the aim of impacting technological innovation process. Mowery (1992) defines innovation policy as one that influences business decisions to develop, commercialize, and implement new technologies.

Hauknes and Wicken (1999b) defined innovation policy in a broader sense, beyond the scope of announced innovation policies (i.e. the strong impact of policy on the implementation of innovation). They include industrial policy, financial policy, trade policy, regulatory and legal measures, and many other policy areas. According to these authors, the policy of innovation consists of both disclosure policies and default policies. The default policy creates an environment and necessary conditions for the issued innovation policy. Edquist (2006) argued that the innovation policy should be drafted and enacted by state intervention, which can lead to technical changes and innovation. These results comprise R&D, technology, infrastructure, regional, and education policies. Hence, innovation policy goes beyond the scope of S&T policy.

Robin Cowan and Gert van de Paal (2000) identified innovation policy as a set of policy actions that can be carried out to increase the quantity and effectiveness of innovation activities. Here, innovation activities refer to the creativity, adaptation, and dissemination of new or improved products, processes, or services.

The European Commission on innovation argues that innovation policy is not merely focused on R&D, but its main objective is to plan the best measures to promote a favorable environment for innovation, and to spread knowledge and technology in the system. A favorable institutional environment includes demands for innovation, a macroeconomic environment, healthy and effective competition, good S&T links, access to venture capital, professional management for business formation, conditions for forming networks, supportive structures, and educational platforms (European Commission 2002).

All in all, innovation policy can be understood as state interventions that create the favorable conditions and environment to promote the most beneficial economic change, encouraging development of human resources, emerging new ideas and realizing these new ideas into goods/services and processes. Studies focus on linking independent policies such as S&T policies, industrial policies, education policies, trade policies, investment and financial policies, etc. into a systematic set.

From the identification of the model of the STI system, including a set of constituent components that interact in each pair of M-R-D-P-M as shown in Fig. 1.3, it is possible that STI policy promotes the interaction between these constituent elements.

According to Vu Cao Dam, there should first be a policy to influence the Market (M) element, so that M stimulates the activities of the Research (R), Deployment (D), and Production (P) elements. First of all, the origin of innovation must stem from the M element. Enterprises must constantly improve their competitiveness in the market. That is the reason for the existence of the business. Competitiveness in M must be reflected in the P capacity. Production capacity must be created through technological innovation capabilities. Then technological innovation may stem from R&D or technology transfer. That is the way to innovate according to the market-pull policy.

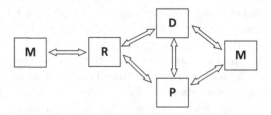

Fig. 1.3 Interactive relationship of constituent elements in the STI system. (Source: Vu Cao Dam, Lecture on S&T policy analysis, Faculty of Science and Technology, 2015, Faculty of Management Science, University of Social Sciences and Humanities)

There is another method for technological innovation. It is one that stems from the will of the manager who proactively sets the direction of technological innovation research. It is an innovation policy in the trajectory of S&T push, one that is S&T-driven. The manager initiates R&D directions ahead of the market or implements a technology transfer policy, to innovate technology (or a product) and push it into production.

An important part of STI policy needs to be emphasized: human, information, financial, and social resources (social capital).

Human resources include people with innovative knowledge and skills working in all the M, R, D, and P areas. They are linked by innovation goals and specific innovation projects. The way of linking is not necessarily in the administrative framework of formal organizations, but possibly in the form of informal organizations or in the framework of contracts.

Information resources include very different information systems—from patented information, useful solutions information, and information on solutions for product, organizational, managerial, and market innovation.

Social resources include social networks, reputations, brands, and traditions. Brokerage organizations are also becoming an important component of social resources. The importance of social resources is increasingly realized. Therefore, people are more interested in developing policies to maximize social resources.

Through a double interaction between the pairs: M ⇔ P, P ⇔ R, R ⇔ D, D ⇔ P, M ⇔ R, and M ⇔ D as shown in Fig. 1.3, it can be assumed that Market suggests production; Production suggests research; Research suggests deployment; Deployment suggests better production, and vice versa.

The STI policy is the policy of developing those interactions, and that is the foundation of the theoretical basis for the STI system. This is considered the theoretical basis for STI policy analysis presented in the following sections. The STI policy is both a government policy as well as a policy of economic and social organizations or universities, S&T organizations, and enterprises.

- *Method of Impact of STI Policy*

According to Vu Cao Dam, the structure of a policy includes the philosophy/perspective/standard system and conceptual system, as shown in Fig. 1.4.

Fig. 1.4 Structure
(Paradigma) of policy.
(Source: Vu Cao Dam
2011, Policy Science
Curriculum, Faculty of
Management Science,
University of Social
Sciences and
Humanities)

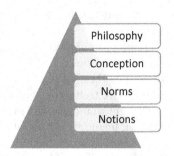

A policy has three stages of impact. According to system theory, the three integration stages of a policy paradigm lead to a transformation in the social paradigm. Similarly, STI policy shows these characteristics, whereby the integration process is a transformation process of the Vietnamese scientific community. That process can take place in three stages as follows[13]:

Stage 1: Destructuration: Breaking the structure of the old paradigm of society.
Stage 2: Restructuration: Establishing a new paradigm of society—i.e. a new structure, completely without returning to the old structure.

There are three ways to restructure. To describe these three ways, it is proposed to refer to Vietnam's paradigm as PV. The world paradigm is PT. A model of compromise is PC. The three ways of restructuring can appear in three integration scenarios as follows:

- *Scenario 1*: When PV is taken as a frame, PT develops in the PV frame. This means that the international one must develop under Vietnam's STI model. But Vietnam has not, in fact, confirmed any models. Therefore, this scenario is less likely to appear.
- *Scenario 2*: When PT is taken as a frame, PV develops in the PT frame. The nature of this scenario is to take the world's development model as a model for Vietnam to follow.
- *Scenario 3*: When PC is taken as a frame, PV and PT develop together in the PC frame. PC compromise scenarios are emerging in a series of countries that are inherently socialist but are not yet successful anywhere.

From the analyzed results of the three scenarios above, Scenario 2 can be affirmed as the only one for the integration process.

Stage 3: Acculturation: The period when policy has completely come to life. The policy paradigm blends into the social paradigm and becomes a single entity growing together.

The impact of the policy is analyzed in accordance with the policy analysis matrix. With the STI integration policy, there will definitely be positive, negative, and peripheral impacts, as well as positive, negative, and peripheral outcomes and implications when taking into account the time series effects.

The InnTeam team, presenting at an innovation seminar at the International Institute of Applied Systems Analysis in Laxenburg, said that innovation policy must aim to create new products, apply new technologies, as well as master and strengthen status in new markets. A trade policy is required to create all of the aforementioned changes and the formation of a dynamic organizational system in all economic and social environments, including economic management and state administration organizations.

Creating innovation and counteracting stagnation in socioeconomic activities are difficult issues for any economy, even stable ones, which are less prone to major fluctuations of the international economic environment. From this perspective, the problem of innovation is more difficult, because the global economy is always subject to fluctuations.

Innovation cannot be seen as an integral part of the economic and social process. The outcome and effectiveness of innovation have profound effects on the pace of economic development and social progress, whereas all economic and social policies and activities have a decisive effect on promoting or restraining the process of innovation.

Therefore, the analysis of innovation policy must be based on an overall approach derived from the needs of life and production, taking it as a basis for identifying programs of research and production as well as supply to consumers. This cycle can be summarized in the following relationship:

M - R - D - P - M

M: market, an expression of social needs
R: scientific research
D: experimental implementation
P: production

The above relationships mean that the study of demand (market) is the basis for deciding scientific research programs. Continuing the period of scientific research is the stage of implementing experimental techniques, trial production, application in industrial production, and market consumption.

In reality, however, the relationship between scientific research and market not only takes place in the simple sequence mentioned above, but can have direct interactions between each pair of the factors. Figure 1.3 shows that these are direct relationships between scientific research and market; research and deployment, research and production; production and market.

In the aforesaid organic relationships, the effectiveness of innovation is determined based on three standards.

1. *Technology*: Can R&D create new products?
2. *Commerce*: After the technical success, can they be consumed?
3. *Economy*: Does the product consumption result in economic efficiency?

The important issue is that the above stages do not take place as a process, but as a result of every employee's individual effort in his/her organization and are placed in a certain economic and social environment.

There are three groups of factors that play a decisive role in the innovation policy.

1.2.5   *Evaluation of Factors Hindering Innovation*

Detecting and evaluating all factors that hinder innovation is an urgent job. The expert consultation method is used to detect and evaluate the factors hindering innovation. After many rounds of investigation and handling of expert opinions, the InnTeam listed categories, including 700 innovative hindering factors and 400 petitions to eliminate those hindering factors. The 700 factors were analyzed and classified into eight groups, ranked in the following order:

1. *Economic management system and method*

This factor had the most comments, which were aimed at institutions and systems of national economic management sectors. The most significant issues include:

- Concentrated nature of the excessive powers of industry management organizations.
- Lack of binding between centralized management and corporate structure.
- Inflexible system of leadership and management at the level of the entire national economy and corporate level.
- No necessary assessment by society for innovation.
- Economic regulation system of frequent fluctuations.
- Inauthentic pricing and valuing systems.
- Lack of free movement of capital.

2. *Innovative and organizational system of innovation system*

This group of factors belongs to the types of difficulties in R&D, market and production organizations, and includes the following factors:

- *Lack of links in the research cycle*: deployment, production, and consumption.
- Most of the current organizational ways of economic units are not favorable to receive innovation.
- There is no model of flexible industry organization and the establishment of quick application of innovative solutions.
- Business conditions and establishment of businesses are limited.
- Development projects are given arbitrarily due to lack of market knowledge.
- No understanding of consumer needs.
- Innovative activities are intended to satisfy the demands that are imposed subjectively, artificially, not market-oriented. What is common is that technical progress plans are formally formed in businesses and technology research institutes.

3. *Benefits*

The third group includes elements relating to the motivation to innovate, the stimulus to the benefits, and the form of expression of different interests:

- Personal and organizational interests are ignored to do research, production, and consumption.

- Innovation is not considered one of the most important goals of an economic organization.
- Benefits of businesses in stable conditions often conflict with the dynamics of innovation.
- Workers' interests have not been cared for.
- Lack of economic urgency (there is no case of corporate bankruptcy as a result of inefficient operation).
- Insufficiency of social recognition for creative workers.

4. *System and methods of enterprise management*

Major deficiencies usually occur in managing the innovation process and the organizations conducting innovation.

- Replacement of business sense with the thought of accomplishing a simple task.
- Lack of independence and self-responsibility of companies.
- Management and planning levels are very low.
- Setting up an incorrect sequence of innovation goals.
- Lack of follow-up of facilities as well as of the entire national economy to build goals.
- Many obstacles to making decisions to focus forces.

5. *Situation of reserves*

The situation of reserves refers to the quantity and quality of hand-delivery and adjustment of resources (finance, labor, supplies, and equipment) needed for innovation.
Major obstacles include:

- Lack of financial resources for experimental implementation.
- Reserves for innovation are dispersed, not beyond the necessary threshold for application.
- Relative backwardness of workers' qualifications.
- Low education level.

6. *Human resources policy and training*

- Non-continuous selection of human resources. The promotion of human resources from the commune level is delayed, causing a counter-selection.

- Insufficient business-minded managers and inadequacy of production professionals.
- Innovative solutions are sometimes offered arbitrarily by managers who have no knowledge.
- The system of competent training hinders creative ideas.

7. Market situation

Market stimulus and the ability to implement innovation are also among the factors that hinder innovation. The most prominent issue is the lack of competition due to:

- Exclusive position of the factory and commercial organizations.
- Lack of competition in the domestic market, resulting from a scarce economy.

8. Human factor

- Fear of the new.
- Jealousy and envy.
- Delay.

It should be noted that the above comments are not of a few individuals, but of many participants in expert groups. The results of the analysis show that there is no fundamental distinction between the recommendations of experts in the fields of agriculture, industry, state management, and the investigation data-handling experts themselves.

The consulted experts provided a series of organizational and management recommendations to eliminate or reduce the factors that hinder the process of innovation. These recommendations are divided into five groups, ranked by the InnTeam team in descending order of importance:

(a) *Development of forms to expand enterprise autonomy in production and business*

Creating flexible organizational forms to allow institutions to overcome "administrative class order" is not suitable for the nature of economic relations. It is necessary to develop legal frameworks for the activities of enterprises, cooperatives, and other economic organizations on the principle of special assurance of business activities and capital mobilization among people.

(b) *Simplification and unification of economic regulatory institutions*

This pertains to reviewing the principles of forming corporate funds, which ensures enterprises of the freedom to use financial resources brought back by business activities. It is also necessary to identify a competitive price policy that reflects the market situation.

From the current market economy perspective, these proposals are too humble. It is quite similar to the market economy system in Vietnam, where the state still plays a decisive role in a series of relations between enterprises and the market.

(c) *Formation of benefit system to stimulate innovation activities*

The basic principle of the benefit system is to ensure the proper integration of personal interests, corporate interests, and the benefits of the entire national economy. The benefit system ensures the combination of spiritual and material encouragement, creating a driving force for the innovation process.

(d) *Creation of creative atmosphere in society*

There needs to be awareness of the importance of innovation at all levels of management in institutions to encourage creativity in society. It is necessary to select human resources that can create trust among workers and encourage interests in creative activities. Promoting a creative society also requires the support of mass media, and schools' curriculum should be aligned with this objective.

(e) *Improvement of economic regulation institution*

The analyses suggest that this is a flexible application of the law of commodity production and of market institutions in a more conscious manner that can reduce bureaucracy and continue to expand democracy in the economy.

Innovation is a matter of intimacy for every production platform. Without innovation in human history, today's civilization would not exist.

Therefore, innovation is never a self-operation; it is determined by the capacity and excitement of human creativity, as well as the various forms of collective labor and the socioeconomic environment. All these factors are dependent on the modality of the economic regulation, a regulating modality that creates high-inertia production by trail or dynamic production, often innovating toward efficiency. As such, the following points are considered:

First, the problems of innovation have long been of concern in the country's social and economic system.

Second, integration issues have also been of interest since many decades ago.

Third, the superiority of the socialist economic and social organization system has given way to more openness.

Fourth, there needs to be recognition of some of the limitations in the social and economic system of the socialist state.

1.2.6 R&D Activities

R&D activities are an important part of S&T. These are systematic and creative activities carried out to increase the amount of knowledge, including human, cultural, and social knowledge. Knowledge is used to create new applications.[14] Thus, R&D is a part, but not all, of S&T activities. According to the United Nations Educational, Scientific and Cultural Organization (UNESCO) (1984) and the OECD (2002), the basic elements for determining R&D activities include:

- creativity;
- novelty;
- use of scientific methods; and
- creating new knowledge.

R&D statistics have been gathered in many countries for more than half a century. In 2012, the OECD celebrated the 50th anniversary of the publication, *Standard Practice for Survey of Research and Development*, commonly known as the Frascati Handbook.[15]

The Science and Technology Law of Vietnam in 2013 identified R&D activities including:

- *Scientific research*: Discovering, detecting, and understanding the nature and laws of things, natural phenomena, society and thinking;

the creation of solutions to apply in practice. It includes *basic research* (a research activity to discover nature, the rules of things, natural phenomena, society, and thought) and *applied research* (a research activity that utilizes scientific research findings to create new technologies and technological innovations to serve the interests of people and society).

- *Technology development*: The use of results of basic and/or applied research through experimental implementation and trial production to improve existing technology and to create new technology. Technology development includes *experimental deployment* (application of R&D results to create new technology products in the form of samples) and *empirical production* (application of experimental implementation results to test production to complete the new technology and new products before they are put into production and life).

In UNESCO's view of S&T activities, *technology development* includes: (1) technology extension (extensive development) after technology has been mastered in production, and (2) technology upgrade (intensive development or upgrading of technology). Thus, technology development will be the stage after the *deployment* phase.

According to the OECD and UNESCO, R&D consists of fundamental research, as well as applied research and development (experimental development). The development phase comprises the prototype, which is piloted to build technology, following which a '0' series trial production is conducted to confirm the stability and reliability of the production technology by the pilot-generated prototype.

1.2.7   *International S&T Integration*

International integration is taking place in the context of increasing globalization. By recognizing the inevitability of international integration for development, most countries expand international cooperation relations toward extensive international integration.

The general definition of *international integration*, according to Mai Ha (2015), is as follows: *International integration is the process of developing and integrating to become an active component of the international system with unified institutions, ensuring long-term benefits for the involved parties.* In this concept, international integration has five characteristics:

voluntariness, acceptance of common rules, conformity, fair competition, and sustainable benefits.

International S&T integration is thus understood as: *The process of national S&T development and integration to become an active component of the international S&T system with unified institutions, ensuring long-term benefits for countries and scientific communities.*[16] International S&T integration also carries the five characteristics of international integration aforementioned.

Currently, international S&T integration has become an indispensable element in S&T development policies in each country. Mai Ha introduced some main forms of S&T integration[17]:

(a) Coordinate scientific research and technological development activities globally or in areas that address one or more sets of issues within a certain period of time. This kind of coordination is often implemented under the principles of promoting the advantages of each country and rationalizing the common goal to achieve the highest efficiency. Examples may include a joint research project on the environment in the Mekong Delta, or a general research project on the Ebola virus and human immunodeficiency viruses (HIV).

(b) Participate in international forums as an international forum member; fully, actively, and proactively participating in S&T activities. This also refers to organizing S&T forums based on the principles of openness and equality, where participating countries must abide by common institutional regulations and standards. Some examples may include the Globelics Forum, Asialics, International Air and Space Association Forum.

(c) S&T activities are integrated on the basis of the international integration of countries. Intergovernmental committees periodically hold conferences and coordinate S&T activities according to the strategy and general development orientation of the integration community under the world's common standards, ensuring intellectual property rights for the results of scientific research and technological development, as well as contributing resources and sharing benefits under jointly agreed commitments. Examples include ASEAN Commission of Science, Technology and Innovation (ASEAN COSTI) and APEC Industrial Science and Technology Working Group (APEC ISTWG).

(d) *National S&T development*: This requires countries, organizations, and individuals to cooperate and associate to conduct joint research to mobilize more resources, shorten time, reduce costs, and, more importantly, attract S&T human resources.

(e) *Enhance the role and influence*: International and regional organizations implement S&T cooperation policies to help developing countries and enhance their role and influence.

(f) *Expand production, markets, and increase profits*: Developed countries and transnational companies expand cooperation to maintain and strengthen their monopoly in expanding production, markets, and increasing profits.

(g) *Enhance competitiveness*: Foreign investors train S&T human resources and skilled workers as well as conduct technology transfer to the host country. Developing countries must prepare human resources, strengthen materials and technical facilities for research and training, improve information and telecommunications infrastructure, and adjust legal regulations.

The trend of international S&T integration often stems from the establishment of international S&T links to professional fields such as electronic equipment, informatics and telecommunications, chemicals and transportation equipment, transportation equipment, space research, ocean, environment and other high-tech fields.

For developing countries, international S&T integration is a trend that helps promote their domestic S&T activities. This allows them to effectively exploit the world's S&T achievements as well as attract foreign human resources and technology to improve and develop their domestic S&T capacity. This contributes to the implementation of the strategic objectives of socioeconomic development and step-by-step integration into the world's knowledge economy. Vietnam's S&T development is an integral part of its socioeconomic system. The country needs to mobilize and utilize the strong impacts of major trends in the world and the region, especially in the context of the integration of the global knowledge economy. In the context of domestic and international socioeconomic conditions, Vietnam will face numerous challenges and must quickly grasp opportunities to develop and integrate into the international S&T arena.

After integrating into the international S&T system, countries can enjoy the following benefits:

- *Ability to access information quickly and objectively:* In the context of integration, countries are more likely to access quickly, fully, and objectively the world's modern achievements, receive the transfer of appropriate technologies that are not self-sufficient domestically. This can increase the domestic technology capability, develop new industries to serve socioeconomic development, and allow industrialization and modernization processes to emerge through shortened paths.
- *Possibility of international cooperation to enhance S&T potential:* In the context of low S&T level in the country, expanding international cooperation in the field of training and research allows people to pursue training and conduct research in developed countries. This can improve the level of S&T human resources and supplement resources (in terms of knowledge, finance, and information), which can quickly build domestic S&T potential.
- *Ability to shorten distances and develop quickly and sustainably:* In the global economy and international integration (i.e. countries see fierce competition in terms of the economy, S&T, and the requirements of compatibility with international practices), Vietnam, with its traditional fondness for learning, is able to shorten the development gap compared to other countries in the region and the world. But that is only the necessary condition. In order to achieve that, the country needs systematic innovation and be able to create a healthy and dynamic development environment for creative thinking and for all labor forces.

1.3 Approaches and Theories of Applied Research

1.3.1 Research Approaches

1.3.1.1 System Approach

STI is not an isolated system, but it is always enacted in a changing S&T environment that is increasingly integrated. The socioeconomic environment is increasingly unpredictable under the effects of policies and diplomatic relations. Therefore, STI must be considered a system with separate components and functions that are closely linked and interrelated. Moreover, when considering the STI system, it is not possible to set this system separately from other systems. It is necessary to create internal links within the organization and ensure harmony in the interrelationships of the system with the environment.

1.3.1.2 Interdisciplinary Approach

When studying the STI system, we need to apply the knowledge and methods of multidisciplinary sciences to ensure that the system is reviewed in detail and fully. From the perspectives of academic managers, economists, sociologists, and psychologists, STI systems are identified as a whole, with interactions that produce great influence on other factors in the economy and society. An economist aims to maximize profits from an invention, sociologists consider the impact of such an invention, psychologists consider people's responses to the invention, a business manager is responsible for directing and developing the direction for the invention to satisfy all the economic, social, and psychological conditions.

1.3.1.3 Logical-historical Approach

The logical-historical approach in STI system research allows one to see the overall appearance, development, and evolution of the system's activities, and at the same time identify the development cycle of an STI system as well as that of an STI policy. Based on the development trend of history combined with empirical research, researchers can make predictions for new development trends of the STI system. Collectively, appropriate measures and activities can be designed to build and develop STI systems. In addition, through studying experience in the development of STI systems, researchers can devise new ways to prevent and avoid repeating existing mistakes. For example, the S&T revolution that has taken place in the world since the 1980s is the fourth S&T revolution (the preceding three being the industrial revolution in the seventeenth century, the first industrial revolution in the eighteenth and nineteenth centuries, and the second industrial revolution in the late nineteenth and early twentieth centuries), and it will continue to upset the development of the foundation of the world's economic sectors. Those that can tolerate the high flow of knowledge will increasingly occupy an important position and determine the general economic development trend of the countries.

1.3.1.4 Top Down-Bottom Up Approach

The STI system should be examined in two ways: top down and bottom up. That is the direction of STI system management, and proper attention must be paid to the direction from the STI components to the management system. The compatibility of two directions is very important—it requires legal corridors and multidimensional relations in each activity of the system to be available.

1.3.1.5 Globalization Trend Approach

Globalization is a term that appeared in the last decades of the twenty-first century. This is an important trend that greatly affects the ideology and policies of organizations in the twenty-first century. According to the OECD: *Globalization is the process that takes place due to the change in technology, long-term continuous growth in foreign investment and international resources and globally large-scale formation in new form of national linkages among companies and countries and the change in global competition nature.*[18]

The trend of globalization and international economic integration is increasing. This is both a process of cooperation for development and a process of struggle between countries to protect national interests. In order to survive and develop in an increasingly competitive environment, the requirements for increasing labor productivity, frequent renovation and improvement of product quality, technological innovation, and innovation of organizational management modes, are growing severe. In particular, in the context of economic globalization, the great achievements of information technology—communication, the popularity of the internet, as well as the development of e-commerce, electronic business, electronic banking, and e-government—are creating new competitive advantages for countries and businesses.

If developing countries are not actively training human resources, strengthening the information-telecommunications infrastructure, or adjusting legal regulations, the risk of lagging behind is high and they will inevitably suffer losses in international exchange relations.

1.3.2 Theories of Research Use

1.3.2.1 Theory of National Innovation System, Regional Innovation System, Industry Innovation System

Theories on the national innovation system (NIS), regional innovation system, and sectorial innovation system help to provide the theoretical foundation of the STI system. The purpose of examining the STI system in line organization is to look at it as a whole where different levels are associated, with different influences.

1.3.2.2 Theory of Innovation Cluster

An innovation cluster is understood to be a group of businesses, research centers, and investors in a geographical region, working together to create new products, new technologies, and new businesses (Bortagaray and Tiffin 2000). In order to form an innovation cluster, the following three groups of requirements need to be satisfied:

- There must be entities, with relationships among the elements.
- The elements must form a whole, where each element must have its own function.
- There must be a distinction among elements inside or outside clusters.

The concept of innovation clusters is focused on the innovation process at the local/regional level. However, the innovation cluster can go beyond administrative boundaries, while the concept of innovation system at the national level is based on administrative boundaries. During a discussion about research projects on innovation, Hoang Xuan Long (2013) presented a summary of the similarities and differences between the NIS, the subnational innovation system, and the innovation cluster. Thus, the innovation cluster shares many basic characteristics with the NIS approach. They all focus on the interaction and interdependence of the elements of the system and the impact of those relationships on the efficiency of innovation and economic efficiency. Basically, the innovation cluster is seen as a miniature NIS. In the innovation cluster model (as well as the NIS in general), the key elements of interaction are knowledge, finance, and people. The main activities are to create knowledge, as well as transfer and use knowledge. To achieve these goals, innovation agents interact with each other in the transfer of knowledge, finance, and human resources.

Cooperation between the elements of the innovation cluster (including organizations such as businesses, research institutes, universities, as well as managers and intermediary organizations) that serve the innovation processes is an important factor that requires clarification in the framework of innovative cluster analysis.

According to Nguyen Thi Minh Nga (2006), features of the innovation cluster include the following:

- Innovation enterprise is in the center of the cluster.
- Suppliers of raw materials are partners in the cluster.
- R&D organization is divided into two groups: (i) research institutes specializing in the field of clusters; and (ii) research institutes not

related to the field of clustering but implement a number of related technology fields.

- Knowledge-intensive business services including design companies and technical consulting companies.
- Business services, which are indirect support factors for the innovation process (tax, banking, law, marketing, etc.).
- *Universities*: University researchers specialize in a small field of cluster or other related fields, and they link with innovative businesses through research contracts between researchers/school teachers and business engineers.
- The government functions as a donor for R&D as well as promotes technology transfer and innovation activities by forming R&D and technology programs to encourage research.
- Supportive organizations (associations, committees, project management offices, and technology licensing organizations).

Collaborative relations between the elements of the innovation cluster will arise through the three flows of knowledge, finance, and human resources.

First, cooperation between businesses is one of the most significant links in the innovation cluster.

Second, in the relations between enterprises and R&D organizations, enterprises can exploit the research infrastructure systems of R&D organizations to serve their innovation activities. Government-funded research institutes and universities provide not only basic knowledge for industry but also a new source of valuable methods, tools, and skills. Enterprises provide financial support and research areas for R&D organizations.

It should be noted that there are three levels of association:

- There is no association.
- There are associations but not for clusters.
- There are associations in the cluster.

In the process of referring to each specific context, it is necessary to clarify which category of the associated cooperation of the survey will be in the categories listed below. Then researchers will comment on each case by following an inductive approach.

- Random/systematic (this is the cluster association).
- Bilateral/multilateral.

- With support from the state/without support from the state.
- *Specific forms of association*: Contract; borrowing experts; visiting and surveying competitive information from competitors.

1.3.2.3 Theory of S&T Push, Market Pull

S&T Push

For Michael J.C. Martin (1994), S&T push is understood to be a research result, where technology is put into production and consumed in the market without taking into account the purpose of enterprise existence or consumer demands. It is a proactive policy of pushing S&T into production and life. The policy has an outstanding value during a war, when governments want to take advantage of all S&T achievements to create weapons. This philosophy has lasted since the industrial revolution (seventeenth–eighteenth century) and was most prominent in the 1950s and 1960s, especially during the period of economic recovery after the Second World War.

Based on the literature reviews by Martin (1994) and Vu Cao Dam (2011) on S&T push policy, it is found that the S&T push model has some consequences, such as research and implementation that are not based on market needs. This results in the product being rejected by the market because it fails to meet consumers' needs. This also affects funding for reinvestment in S&T activities.

Technology Push and Market Pull

In the process of innovation, market pull is of special importance. A lot of time and money can be invested in R&D work to develop products, but if one does not see clearly the needs of the market, then even inventions of special significance will not be accepted.

According to Vu Cao Dam (2011), the market-pull policy is a development policy in the context of an economic system transforming into an open market. The market will attract S&T, serving the objectives of cooperation and competition. This policy has been in practice since the 1980s and 1990s.

Before implementing a market-pull policy, consumer and market demands should be surveyed first. Then R&D is carried out according to the demands of the market, so that its products will be consumed. Otherwise, research results would be rendered useless. The market consumes the results of R&D, and this leads to the expansion of funding for R&D reinvestment, increasing efficiency in S&T activities.

1.4   METHODS OF INFORMATION COLLECTION AND PROCESSING

The data for this study was collected based on both qualitative and quantitative methods so that the author could assess the current status of each component of the STI system and assess Vietnam's STI policy.

1.4.1   Methods of Information Collection

1.4.1.1   Qualitative Investigation Method

Qualitative research was designed based on the analysis of policies directly and indirectly relating to STI activities issued by the government, the National Assembly, as well as the Ministry of Science and Technology and the provincial Departments of Science and Technology. At the same time, the author also looked at the analysis of development strategies and the socioeconomic situation of Vietnam. Qualitative research is known to be based on the following methods: document research, seminars/talks, observations, in-depth interviews, and group discussions.

Document Research

The analysis of domestic and foreign sources and data included a thorough desk study (document collection, document analysis, and document synthesis) from the OECD, UNESCO, United Nations Development Programme, European Union sources, international journals, Vietnamese magazines, and S&T white papers. The synthesis method combined with analysis was done based on an analysis, evaluation, and synthesis of the documents relating to the evaluation of the STI system and the development of NIS. The author also applied the survey data from the two state-level projects chaired by the author, namely Projects KX06.06/11-15 and KX01.01/16-20. The procedures and techniques for data collection are described below.

In-depth Interviews

In-depth interviews are conducted with policymakers and implementers, experts on innovation and S&T management, and representatives of S&T organizations (institutes, schools, businesses). The content of in-depth interviews is different for each category.

- *Primary object group*: Policymakers and policy implementers, and five experts on creative innovation and S&T management. The interviews revolved around the development trend of STI policy today, evaluating the factors that impact STI policy, identifying Vietnam's status of STI, and predicting future growth trends.
- *Secondary object group*: Representatives of S&T organizations (including 10 representatives from universities, five from research institutes, and 25 from enterprises). The in-depth interviews with these participants focused on STI activities in their organizations and the policy orientation for STI activities proposed by their organizational leaders. The interviews also aimed to assess factors affecting their organizations' STI activities, and the disjuncture between the organization and elements in the NIS.

Group Discussion

Group discussions were conducted for individual groups who organized STI activities in their organizations. These participants included lecturers, researchers, experts, workers at enterprises, and other workers. This method was carried out in two groups (each group comprised two or three individuals/organizational representatives) in each type of S&T organization. The discussions focused on the STI activities in their organizations, the effectiveness and benefits of STI activities, difficulties, barriers when they were assigned to conduct STI activities, policies that affected the organizations' STI activities, and the role of the STI activities for the development of individuals and the organization.

Seminars and Roundtable Dialogues

Seminars and roundtable dialogues were organized for groups of 10 to 30 experts during the process of implementing research issues. The topics of these activities included:

- Detecting common and different points of STI activities in each type of organization (universities, institutes, and enterprises).
- Identifying and assessing barriers to STI activities at S&T organizations.
- Developing a framework for assessing STI activities at S&T organizations.

- During the implementation process, the author organized seminars and talks with participation and comments from many experts on S&T management, leaders of the Ministry of Science and Technology, Department of Science and Technology, leaders of universities, research institutes, lecturers, and Master's students.

Observations

The research team conducted field observations at two academies (Vietnam Academy of Science and Technology and Vietnam Academy of Social Sciences), four universities (Vietnam National University, Hanoi (VNU), Vietnam National University Ho Chi Minh City (VNUHCM), Hanoi University of Science and Technology, and Thai Nguyen University), and 20 enterprises (including 12 S&T enterprises). The field trips enabled the research team to observe real situations and the use of resources for STI activities at each type of S&T organization in the NIS.

1.4.1.2 Quantitative Investigation Method
This was the main research method that the research used to investigate and explore STI activities at the S&T organizations. The author used two sample questionnaires:

Questionnaire for Organizations (Large Enterprises/SMEs, Universities, and Research Institutes)

The questionnaire aimed to provide an overview of the current status of STI activities in S&T organizations, analyze resources for STI activities, identify difficulties in STI activities at S&T organizations, as well as assess the impact of STI policies on the activities and development of the organizations and the relationship between the elements in the NIS.
The questionnaire had three main sections:

- general information about the organization (type, legal model, and time of establishment);
- resources for STI activities (human resources, finance, and facilities); and
- operational results (number of topics/projects at home and abroad, number of domestic and international inventions, and publications).

Questionnaire for Individuals Working at Organizations (Institutes/ Schools/Businesses)

This group included those who were participating in STI activities at the S&T organizations where the author had conducted the surveys. The author distributed the questionnaire to 224 individuals. The questionnaire gathered data on the following:

- general information (working time, qualifications);
- personal assessment of the STI system of organizations and of Vietnam (working environment, importance of policy, environment, and resources for STI activities, as well as individual need for STI activities); and
- individual recommendations.

The purpose of the questionnaire was to examine the outlook of the participants who directly engaged in STI and were directly affected by STI policies. This was the most important premise for proposing STI policy implications later.

The research team used a non-random sampling technique. This method is used when the specific list of the overall population cannot be defined (a large number of S&T organizations were located in big cities like Hanoi and Ho Chi Minh City).

Sample structure:

Total number of questionnaires distributed

Questionnaire type	Number	Percentage
Individuals	224	57
Organizations	169	43
Total	393	100

Questionnaires for organizations (institute/school/business)

Questionnaire for organizations	Number	Percentage
Research institutes	60	35.5
Universities	5	3
Enterprises	104	61.5
Total	169	100

Questionnaires for individuals working at S&T organizations (institutes/schools/businesses)

Questionnaire for each organization	Number	Percentage
Research institutes	120	53.6
Enterprises	104	46.4
Total	224	100

Specific sample structure of personal sheets:

The demographic information of the 120 individual participants working at research institutes:

		Number	Percentage
Gender	Male	72	60
	Female	48	40
	Total	120	100
Academic title, degree	Bachelor	14	11.6
	Master	25	20.8
	Doctor	60	50
	Assoc. Prof.	21	17.5
	Total	120	100

The demographic information of the 104 individuals working at enterprises:

		Number	Percentage
Gender	Male	81	77.9
	Female	23	22.1
	Total	104	100
Age	Under 30 years old	12	11.5
	30–45 years old	78	72.2
	Over 45 years old	17	16.3
	Total	104	100
Work title	Enterprise level leader/manager	36	34.6
	Departmental level leader/manager	47	45.2
	Group level leader/manager	3	2.9
	Office staff	6	5.8
	Technician	12	11.5
	Total	104	100

The author also conducted another sociological survey to separately investigate the status of S&T quality sources in the four major S&T organizations: Vietnam Academy of Science and Technology, Vietnam Academy of Social Sciences, Vietnam National University, Hanoi (VNU), and Vietnam National University Ho Chi Minh City (VNUHCM), as shown below:

Total sheets

Questionnaire type	Number	Percentage
Personal	1268	95.2
Organizational	64	4.8
Total	1332	100

Organizational sheets

Organizational sheets	Quantity	Ratio
Vietnam National University, Hanoi	10	15.6
Vietnam National University Ho Chi Minh City	9	14
Vietnam Academy of Science and Technology	25	39.1
Vietnam Academy of Social Sciences	20	31.3
Total	64	100

Personal sheets working at the four survey units

Individual sheets of each organization	Quantity	Ratio
Vietnam National University, Hanoi	261	20.6
Vietnam National University Ho Chi Minh City	352	27.8
Vietnam Academy of Science and Technology	324	25.6
Vietnam Academy of Social Sciences	331	26
Total	1268	100

For this object group, the questionnaire was conducted through one of the following forms: distributing the questionnaires directly to the informants, sending the questionnaire via email, or via telephone.

1.4.2   Methods of Information Processing

1.4.2.1 Processing Information by SPSS

The primary and secondary data were processed via SPSS program version 16.0. This was done through the following steps:

Step 1: *Data encryption*: Qualitative data (qualitative variables such as the type of variables of properties, superior or inferior) was converted (coded) into numbers. Quantitative data (quantitative variables are usually represented by numbers and these numbers can be in the form of continuous variations such as age, capital call amount, space, etc.) did not require encryption.

Step 2: *Data entry*: The data was entered and stored into the data file.

Step 3: *Correction*: Checking and detecting errors in the process of entering data from the handwritten data sheet into the data file on the computer.

After the data had been encoded and processed, the team conducted data analysis to determine the research issues. The research team compared data sets in the same class, for example, making comparison in a group with positive or negative data, the percentage, pairing, measuring differences or similarities between variables, so that a comparison between two or three groups was conducted against the same criteria. In addition, the research team applied the efficiency of data analysis according to the average score (mean) of each factor to determine the value level and ranking factors from which to make objective, scientific comments and conclusions.

1.4.2.2 In-depth Interviews

The author conducted in-depth interviews with 40 representatives of the S&T organizations and five S&T experts and managers to find out more about barriers in the current STI policy as well as difficulties and achievements of STI activities at S&T organizations. These interviews were integrated and some of the quotes are cited in this book, especially those related to the reality of the organizations' STI activities, the status of STI policies, and the orientation of STI policy solutions in Vietnam in the future.

NOTES

1. The expression is used by Hoang Phuong, published in VnExpress Newspaper. Available at https://vnexpress.net/tin-tuc/su-kien/30-nam-doi-moi/cong-cuoc-sap-xep-lai-giang-son-truoc-dai-hoi-doi-moi-1986-3510505.html
2. Charles Edquist, Systems of Innovation: Perspectives and Challenges, The Oxford Handbook of Innovation, available at http://www.oxfordhand-

books.com/view/10.1093/oxfordhb/Medi99286805.001.0001/
oxfordhb-Comic99286805-e-7 # oxfordhb-Comic99286805-div1-41

3. Original: "An innovation is the implementation of a new or significantly improved product (good or service), or process, a new marketing method, or a new organisational method in business practices, workplace organisation or external relations", OECD, 2005, "The measurement of scientific and technological activities: guidelines for collecting and interpreting innovation data: oslo manual, third edition" prepared by the working party of national experts on scientific and technology indicators, OECD, Paris, p. 146.

4. Crossan, M., and Apaydin, M. (2009). A Multi-Dimensional Framework of Organizational Innovation: A Systematic Review of the Literature. *Journal of Management Studies*, 47(6), pp. 1154–1191.

5. Original: "New or Changed Object, Realizing or Redistributing Value", available at https://www.iso.org/obp/ui/#iso:std:iso:9000:ed-4:v1:en

6. Report of the 1992–1995 Fiscal SAREC Project, NISTPASS Institute Library Document.

7. View http://www.encyclopedia.com/topic/Nguyen_Van_Linh.aspx

8. View http://vi.oldict.com/renovation/

9. Detail: "Doi Moi (English: *Renovation*) is the name given to the economic *reforms* initiated in Vietnam in 1986 with the goal of creating a 'socialist-oriented market economy'. The term Doi Moi itself is a general term with wide use in the Vietnamese language, however the Doi Moi Policy refers specifically to these reforms".

10. See https://en.wikipedia.org/wiki/Meiji_Restoration

11. Carlsson, B. and Stankiewicz, R., On the Nature, Function, and Composition of Technological systems, *Journal of Evolutionary Economics* 1(1991) 93–118.

12. Cited from Pham Hong Quat and Nguyen Duc Phuong (2013), *University/Research Institute in STI: The Reality of Knowledge Transfer and Suggestion for Some Basic Solutions*.

13. Vu Cao Dam (2015), *Lecture on S&T Policy Analysis*, 2015, Faculty of Management Science, University of Social Sciences and Humanities

14. UNESCO (1984), *Manual for Statistics on Scientific and Technological Activities*, ST.84/WS/12. Paris: UNESCO.

15. OECD (2002), *Proposed Standard Practice for Survey of Research and Development—Fracasti Manua*. Paris.

16. Mai Ha (2015), International integration of science and technology of Vietnam in the new period, *Vietnam Social Science Review*, No. 7 (92), p. 108.

17. Mai Ha (2015), International integration of science and technology of Vietnam in the new period, *Vietnam Social Science Review*, No. 7 (92), p. 108.

18. OECD (2005), *Handbook on Economic Globalization Indicators*.

REFERENCES

Ab Aziz, K., Harris, H., and Norhashim, M. (2011). University Research, Development & Commercialisation Management: A Malaysian Best Practice Case Study. *World Review of Business Research*, 1(2), pp. 179–192.

Bergek, Anna, Jacobsson, Staffan, Carlsson, Bo, Lindmark, Sven, and Rickne, Annika. (2008). Analyzing the Functional Dynamics of Technological Innovation Systems: A Scheme of Analysis. *Research Policy*, 37(3), pp. 407–429.

Bortagaray, I., and Tiffin, S. (2000). *Innovation Clusters in Latin America*. 4th International Conference on Technology Policy and Innovation Curitiba, August 28–31, Brazil.

Carlsson, B., and Stankiewicz, R. (1991). On the Nature, Function and Composition of Technological Systems. *Journal of Evolutionary Economics*, 1(2), pp. 93–118.

Charles River Associate Ltd. (2003). *Innovation Policies in Singapore, and Applicability to New Zealand*. Wellington: Ministry of Economic Development Wellington.

Cowan, Robin, and van de Paal, Gert. (2000). *Innovation Policy in a Knowledge-Based Economy*. Luxembourg: Commission of the European Communities.

Crossan, M., and Apaydin, M. (2009). A Multi-Dimensional Framework of Organizational Innovation: A Systematic Review of the Literature. *Journal of Management Studies*, 47(6), pp. 1154–1191.

Dang Ngoc Dinh. (2015). Innovation of Science and Technology to Effectively Serve for the Development and International Integration of Vietnam. In: Dao Thanh Truong, ed., *Science, Technology, and Innovation System of Vietnam in the Trend of International Integration*. Hanoi: Thế Giới Publishers.

Dao Thanh Truong. (2016). *Chính sách Khoa học, Công nghệ, và Đổi mới (STI) của Việt Nam trong xu thế hội nhập quốc tế: Thực trạng và giải pháp* [Science, Technology, and Innovation Policies of Vietnam in the Trend of International Integration: Situations and Solutions]. Hanoi: Thế Giới Publishers.

Edison, H., bin Ali, N. and Torkar, R. (2013). Towards Innovation Measurement in the Software Industry. *Journal of Systems and Software*, 86(5), pp. 1390–1407.

Edquist, C. (1997). *Systems of Innovation: Technologies, Institutions and Organizations*. London: Pinter.

Edquist, C. (2001). *The Systems of Innovation Approach and Innovation Policy: An Account of the State of the Art*. Paper presented at the DRUID Conference, Aalborg, Denmark.

Edquist, C. (2006). Systems of Innovation: Perspectives and Challenges. In: J. Fagerberg, D. Mowery, and R. Nelson, ed., *The Oxford Handbook of Innovation*. Oxford: Oxford University Press.

European Commission. (2002). Innovation Tomorrow: Innovation Policy and the Regulatory Framework: Making Innovation an Integral Part of the Broader Structural Agenda. Innovation Papers, No. 28.

Fagerberg, J., Mowery, D. C., and Nelson, R. R. (Editors). (2005). *The Oxford Handbook of Innovation*. Oxford: Oxford University Press.

Foster, R. (1986). *Innovation—The Attacker's Advantage*. New York: McKinsey & Co.

Freeman, C., Clark, J., and Soete, L. L. G. (1982). *Unemployment and Technical Innovation: A Study of Long Waves in Economic Development*. London: Pinter.

Gault, F. (2010). *Innovation Strategies for a Global Economy: Development, Implementation, Measurement, and Management*. Northampton: Edward Elgar.

Hammer, R. (2006). Strategic Innovation: The Engine that Propels Business. *FocalPoint Ventures, Inc.* Available at: https://innovationmanagement.se/wp-content/uploads/pdf/InnovationEngine.pdf.

Hauknes, J., and Wicken, O. (1999a). Innovation Policy in the Post-War Period—Trends and Patterns. Oslo: STEP-Group.

Hauknes, J., and Wicken, O. (1999b). Innovation Policy in the Post-war Period. In: *Integrating Regional and Global Initiatives in the Learning Society*, ICTPI KANSAI' 2002 Conference, August 2002, Kyoto, Japan.

Hekkert, M. P., Suurs, R. A. A., Negro, S. O., Kuhlmann, S., and Smits, R. E. H. M. (2007). Functions of Innovation Systems: A New Approach for Analysing Technological Change. *Technological Forecasting and Social Change*, 74(4), 413–432

Hindle, K., and Yencken, J. (2004). Public Research Commercialisation, Entrepreneurship and New Technology Based Firms: An Integrated Model. *Technovation*, 24(10), pp. 793–803.

Hoang Van Tuyen. (2006). *Research on Development Process of Innovation Policy: International Experience and Suggestions for Vietnam*. Hanoi: National Institute for Science and Technology Policy and Strategies Studies.

Hoang Xuan Long. (2013). *Support to the Development of the S&T Strategic Framework through Innovative Systems Approach and Research Reform Mechanism*. Hanoi: NISTPASS.

Kroll, H., and Liefner, I. (2008). Spin-off Enterprises as a Means of Technology Commercialisation in a Transforming Economy—Evidence from Three Universities in China. *Technovation*, 28(5), pp. 298–313.

Lundvall, B. A. (1992). *National Systems of Innovation: Toward a Theory of Innovation and Interactive Learning*. London: Pinter.

Lundvall, B., Intarakumnerd, P., and Vang, J. (2006). *Asia's Innovation Systems in Transition*. Cheltenham, UK: Edward Elgar.

Mai Ha. (2015). International Integration of Science and Technology of Vietnam in the New Period. *Vietnam Social Science Review*, 7(92), pp. 108–113.

Malerba, F. (2002). "New Challenges for Sectoral Systems of Innovation" in Europe. In: *Industrial Dynamics of the New and Old Economy—Who is Embracing Whom?* DRUID Summer Conference, June 6–8, Copenhagen, Denmark.

Martin, M. (1994). *Managing Innovation and Entrepreneurship in Technology-based Firms*. New York: Wiley.

Möller, K., Rajala, R., and Westerlund, M. (2008). Service Innovation Myopia? A New Recipe for Client-Provider Value Creation. *California Management Review*, 50(3), pp. 31–48

Mowery, D. (1992). The U.S. National Innovation System: Origins and Prospects for Change. *Research Policy*, 21(2), pp. 125–144.

Nelson, R. (1993). *National Innovation Systems*. New York: Oxford University Press.

Nelson, R., and Rosenberg, N. (1993). Technical Innovation and National Systems. In: R. Nelson, ed., *National Innovation Systems: A Comparative Analysis*. New York and Oxford: Oxford University Press.

Nguyen Hong Ha. (2004). *Research on Methods and Process of Science and Technology Policy Evaluation—Case Study: Decree 35/HDBT and Decision 782/ TTg*. Ministry-Level Project, Hanoi.

Nguyen Thanh Ha, Tran Ngoc Ca, and Nguyen Vo Hung. (1992–1995). *Report of the 1992–1995 Fiscal SAREC Project*. NISTPASS Institute Library Document.

Nguyen Thi Minh Nga. (2006). *Innovative Cluster Research: Overview of International Experience and Lessons for Vietnam*. Hanoi: NISTPASS.

Nguyen Van Hoc. (2005). *Assessment of Research and Development Organizations*. Ministry-Level Project, Hanoi.

Nguyen Van Hoc. (2012). *Lecture on National Innovation System*. Hanoi: Faculty of Management Science, VNU-The University of Social Sciences and Humanities Vietnam.

OECD. (2002). *Proposed Standard Practice for Survey of Research and Development—Fracasti Manua*. Paris.

OECD. (2005). *The Measurement of Scientific and Technological Activities: Guidelines for Collecting and Interpreting Innovation Data: Oslo Manual, Third Edition*. Paris: Organisation for Economic Co-operation and Development.

Palmer, D., and Kaplan, S. (2007). A Framework for Strategic Innovation. *Innovation Point*. Available at: http://www.innovation-point.com/ Strategic%20Innovation%20White%20Paper.pdf.

Phung Xuan Nha and Le Quan. (2013). Innovation of Enterprises in Vietnam. *VNU Journal of Science: Economics and Business*, 29(4), pp. 1–11.

Pham Hong Quat and Nguyen Duc Phuong. (2013). *University/Research Institute in STI: The Reality of Knowledge Transfer and Suggestion for Some Basic Solutions*. Hanoi: Thế Giới Publishers.

Pham Ngoc Minh. (2015). Science, Technology and Innovation Activities of Enterprises. In: Dao Thanh Truong, ed., *Science, Technology, and Innovation System of Vietnam in the Trend of International Integration*. Hanoi: Thế Giới Publishers.

Porter, Michael. (1990). The Competitive Advantages of Nations. *Harvard Business Review*, 68(2), pp. 73–91.

Schumpeter, J.A., 1934 (2008), *The Theory of Economic Development: An Inquiry into Profits, Capital, Credit, Interest and the Business Cycle*, translated from the German by Redvers Opie, New Brunswick (U.S.A) and London (U.K.): Transaction Publishers.

Stoneman, P. (1987). *The Economic Analysis of Technology Policy.* Oxford: Clarendon Press.

The International Organization for Standardization. (2015). *New or Changed Object, Realizing or Redistributing Value.* [Online] Iso.org. Available at: https://www.iso.org/obp/ui/#iso:std:iso:9000:ed-4:v1:en.

The Swedish Research Council (VR) and The Swedish Institute for Growth Policy Studies (ITPS). (2004). *Report in Government Research and Innovation Policies in Japan.*

Tran Ngoc Ca. (2006). *Vietnam's Innovation System: Toward a Product Innovation Ecosystem.* Conference on Global Innovation Ecosystem, 8–9 September, Kyoto, Japan.

Tran Ngoc Ca and Nguyen Vo Hung. (2012). *Towards an Innovation System in the Field of Agriculture: The Case of Three Products in Vietnam: Vegetables, Tea, and Shrimp.* Hanoi: Science and Technics Publishing House.

UNESCO. (1984). *Manual for Statistics on Scientific and Technological Activities.* ST.84/WS/12. Paris: UNESCO.

Vietnam Ministry of Science and Technology. (2013). *Vietnam Law on Science and Technology.* Vietnam.

Vu Cao Dam. (2011). *Policy Science Curriculum.* Hanoi: Vietnam National University Publishing House.

Vu Cao Dam. (2015). *Lecture on S&T Policy Analysis, Faculty of Science and Technology.* Hanoi: Faculty of Management Science, University of Social Sciences and Humanities.

CHAPTER 2

Experience in Developing STI System in Some Countries in the Context of International S&T Integration

2.1 STI SYSTEM OF SWEDEN

The current population of Sweden is 10,050,389 and equivalent to 0.13% of the total world population.[1] Considered a high-knowledge economy and a world leader in the business environment, Sweden is one of the pioneering countries in innovation that has successfully developed the innovation system and has the highest expenditure on R&D activities. The Swedish innovation system is formed by three perspectives: the national innovation system, the sectorial innovation system (which means a particular industry sector), and the regional innovation system.

One of the main advantages of Sweden is that the scientist who makes the innovation owns the innovation and there are systems put in place that help the scientist to protect the IP of the discovery and to start a company which can be housed in science parks/incubation centers within the university for its growth. Vinnova as well as other agencies like Innovationsbron, GU Holding, and so on provide start-up funds, early growth funds, and after some years, Venture Cap funds.[2] The miraculous rise of Sweden with the strategic goal of relying on STI has brought great success. There are popular products from Swedish S&T activities which have changed the world market. Some examples are Tetra Pak packaging, three-point safety belts, Skype, Bluetooth, multitouch color screen, portable mobile, ultrasound, zipper, and the pacemaker.

© The Author(s) 2019
D. T. Truong, *Perspectives on Vietnam's Science, Technology, and Innovation Policies*,
https://doi.org/10.1007/978-981-15-0571-3_2

Sweden has developed S&T and is considered a good case study for STI system development. Now, the challenge for Sweden is how to maintain the success and orient new development strategies in the context of globalization. The policies that focus on building and developing the Swedish STI system offer valuable experiences for other countries. Sweden is ranked high on all the ratings for the indicators of innovation capacity. The country has strengthened its position as one of the most competitive economies in the world and an innovative leader in the European Union. With the advantages of its business environment, the preferential loans offered by its banks to SMEs create an environment of innovation that is highly convenient for any individual or organization who wants to start R&D. Sweden has always focused on the investment and development of the STI system, considering it both a goal and a means for national development. We can see that Sweden's R&D investment grade has always been among the leaders in the European Union.

For many consecutive years, Sweden has been placed in the top-five list of the most innovative countries in the Global Innovation Index (GII). Sweden has invested a large proportion of its GDP in scientific research and education. With the growth of the high-tech industry, there are many opportunities for businesses. Sweden is an outstanding model for stability between innovation and development according to its rankings in the GII and the Innovation Union.

In Sweden, every four years, the government prepares a Research and Innovation Bill, which defines long-term and short-term goals as well as operating costs for the STI system over the next four years. In the latest such bill in 2012, the Swedish government stated that its STI policy goal remains to be a strong R&D country by improving research quality and enhancing the contributions of the results of scientific research for socio-economic development through the promotion of the commercialization of research results.[3] In addition, the Swedish government promotes the development of STI policies through budget regulations, general S&T activities regulations, and having board members in organizations and S&T organizations. Besides, having a relatively large number of foundations which act independently from official public policymaking is one of the characteristics of Sweden's STI system.

The Swedish STI system is based on a triple helix model.[4] The central task of these components is to focus on the process of commercialization and business innovation. The non-profit governmental and private sectors play a major role as research financiers (investments of these two sectors for R&D account for only 5% of total R&D spending). In 2011 the country invested 3.37% of its GDP in research and 0.65% of its GDP in innovation and structural change for a total of 4.02%.[5]

The structure of the Swedish STI system is characterized by clarity and the clarification of responsibilities among the components of the system. With the three main functions and the three main components of the STI system, organizations have a very close coordination mechanism.

- Function 1: General policy formulation.

The national government and the local government, together with the National Council, play the major implementer role with the support of the Research Policy Council, the Innovation Policy Council, and the Institute for Development Policy Studies. At the local level, the policy is developed in Regional Development Programs.

- Function 2: Sponsorship and policy support.

All three areas (i.e. research institutes/universities, and the government) contribute to the funding. Funding for basic research is primarily for higher-education institutions by three research councils. Funding for R&D activities not related to defense and security is managed by a range of organizations including VINNOVA (Swedish Governmental Agency for Innovation Systems). VINNOVA's mission is to promote sustainable growth by funding needs driven research and developing effective innovation systems. Innovation systems are an analytical perspective for understanding the dynamic connection between actors, institutions and other determinants of the volumes, directions, results, and impacts of innovation processes.[6] In addition to the public sector, the private sector, which includes industries and private funds, contributes greatly to Sweden's general S&T funds.

Another secondary function undertaken by VINNOVA and the Swedish Research Council is policy support. VINNOVA supports policy issues relating to innovation and R&D activities. The Swedish Research Council serves as an advisor on research policy-related issues.

- Function 3: Performing R&D activities.

This function is performed by universities and research institutes. In Sweden, there are about 16 universities participating in R&D activities.

These institutions focus on research and technological transfer. Despite offering humble contributions to the STI system, these research institutes act as an important bridge between research and application, especially for SMEs. Generally, policies are developed by the government and ministries such as the Ministry of Education and Research, the Ministry of Business, Energy and Communications, and the Ministry of Defense, under the support and advice of the Research Policy Council and the Innovation Policy Council.

Pertaining to the resources used to maintain and develop the STI system, finance plays a significant role. At the end of 2012, the Ministry of Business, Energy and Infrastructure issued Sweden's National Innovation Strategy. At the end of 2013, the national conference with the Ministry of Business, Energy, and Infrastructure and its stakeholders evaluated the realization of the national innovation strategy. It emphasized that government policies promoted technology incubator businesses in Sweden with an annual investment of about 3 million euros for VINNOVA's business incubation programs. This is considered an important part of implementing Sweden's innovation strategy. Business enterprises dominate in Sweden's R&D sector. They play the role of a source of funding and an implementation group for R&D activities. In fact, most of the R&D funding sources in Sweden are from/to business enterprises, and in general, the share of total R&D funding between sectors is very low.

The Swedish Research Council, funded by the Ministry of Education and Research, has funded research across all areas of natural and social sciences, health and education. The largest direct sponsorship program is for individuals and small groups in academic circles. This program is based on a yearly selection, and its applicability is evaluated against scientific criteria such as the quality and feasibility of the project. In addition, there are a few special selections by the government or the council. This content and procedure are quite similar to the system of selection by bidding for scientific research projects in Vietnam.

Moreover, some funds are not entirely established for the research centers. For example, the Knowledge Foundation focuses on research and innovation at higher-education institutions (but not universities or colleges). The Swedish Foundation for Strategic Studies specializes in funding potential research projects on innovation. The Bank of Sweden Tercentenary Foundation specializes in funding social science projects and the humanities.[7]

Compared to other European Union countries, the Swedish public R&D sector lies largely in the academic sector. According to the OECD report in 2016, Sweden has 40 publicly financed universities and colleges, including 10 major universities.[8]

In October 2012, the Swedish Research and Innovation Bill set out the orientations for the STI system, focusing on the use and management of the operational budget for the STI system:

- Increase funding for universities and colleges (up to 25 million euros).
- Sponsor a national center for life science research (SciLifeLab) (about 17 million euros).
- Develop targeted funding programs for basic research in life sciences (about 35 million euros) and other areas of particular importance that strengthen Sweden's future competitiveness (about 48 million euros).
- Approve a special funding program for attracting outstanding international researchers to Swedish universities and a special program for young researchers in Swedish universities (about 20 million euros).
- Strengthen the capacity of the research areas through increased support for RISE companies.

The development and practice of Sweden's STI are unique in structure and performance, setting an example for other countries. Sweden's STI system has the following features:

- It has a clear division between the public and the private sectors in R&D activities, and each region is strongly influenced by several factors in the respective area (major multinational companies such as AstraZeneca, Atlas Copco, Electrolux, SKF, IKEA or Tetra Pak in the private sector, and large research universities in the public sector).
- There is a high degree of cohesion between science and research, and a high degree of autonomy among the factors in the public-sector science system.
- Small ministries are government organizations that formulate and implement policies that are closely related, and they, too, have a high degree of autonomy.
- There is a division of responsibilities in the STI system between the Ministry of Education and Research with the function of solving research and innovation issues in the academic sector, while the

Ministry of Enterprises is responsible for solving research and innovation issues in the private sector.

- Horizontal coordination mainly takes place through informal mechanisms between ministries and organizations under the Ministry.
- There is a hierarchy of responsibilities within the STI system for regional governments (districts).
- Universities are required to develop their own research strategies.
- There is a default division of labor rates: a small number of top universities account for a large proportion of the labor rates in the public research area and the specialized higher-education institutions.
- There is a compelling orientation for universities to have two responsibilities: promote self-study and orient research tasks. In fact, universities face difficulties in becoming intermediaries between research and production.

However, based on the Swedish R&D activity indicators, there is a problem known as the "Swedish paradox". This refers to the incompatibility in public investment for R&D activities. Inputs for STI (including investment costs for R&D, labor, S&T information, etc.) are not commensurate with outputs (measured by patents and scientific publications). The limitation in the commercialization of research results is also considered a weakness in Sweden's STI system. Facing such a problem, Sweden has built an STI system to promote spin-off businesses and start-ups to support SMEs, especially the establishment of high-tech SMEs through the dissemination of research results and the support for SMEs to develop scientific research. Besides, R&D activities in the Swedish business sector are largely restricted to a smaller number of companies and most of them belong to multinational enterprises (MNEs), Swedish MNEs, or foreign-owned firms. In 2017, the highest R&D intensities were recorded in Sweden (3.33%). Besides, this country was at the top of the countries with the highest share of R&D expenditure performed in the business enterprise sector (71%).[9]

According to the Global Innovation Report 2018, Sweden moved down to the third position (after Switzerland and the Netherlands).[10] This country keeps its first position in PCT patent applications by origin and gains a first rank in IP receipts and rule of law. Pupil-teacher ratio, GDP per unit of energy use, ease of getting credit, GERD financed by abroad, FDI inflows, and productivity growth are areas of weakness in Sweden. In 2019, Sweden has improved its education and knowledge diffusion measures[11] and is at the second rank in GII top 10. Although Sweden is one of the most competitive countries in the world, with top grades for innovation capacity, some challenges for this country are high taxes and labor regulations.[12] The Innovation Strategy of Sweden to 2020 is based on three main principles: the best possible conditions for innovation; people, businesses and organizations that work systematically with innovation; implementation of the strategy based on a holistic view (in developed coordination between policy areas and policy levels; in dialogue with actors in industry, the public sector and civil society; in a process of continuous learning).[13]

2.2   STI SYSTEM OF GERMANY

According to the EU Scoreboard Innovation Index 2015, the Federal Republic of Germany was ranked as one of the four countries in the "innovation leader" group,[14] with the implementation of innovation higher than that in the EU on average. In the aspect of "open, excellent and attractive research system", Germany obtained a lower score than the EU on average. But in the aspect of "finance and support", Germany was one of the top four countries with the highest rates of R&D investment from enterprises.

Regarding the global technology index, Germany is ranked ninth among the 47 high-income countries and ranked seventh among the 39 countries in Europe in the GII 2018.[15] With the ability to compete with other countries, Germany remains at the top for its technology and know-how. Along with the United States and Japan, Germany is one of the leading technology producers in the world and a leading technology provider in Europe. In addition to training and good R&D activities, infrastructure is an important factor that helps increase technological advantages, creating the creative power and flexibility of businesses and industries, especially in the SME sector.

The economic strength of Germany comes from a strong export-driven orientation that is based on innovation to improve production performance. However, a lot of debate has occurred because of Germany's below-average economic growth rate in recent years. Germany's government has exempted public studies from tax and reduced expenditure on education.

Germany performs well on indicators of human resources in science and technology (HRST). In 2013, the proportion of scientists and engineers among those working in S&T industries was 21.5% in the EU. In the group of experts, Ireland (45.3%), Finland (44.4%), and Germany (43.9%) had the highest number of scientists and engineers. Six member countries including Germany, Austria, France, Italy, the Czech Republic, and Slovakia accounted for about 50% of S&T human resources in the EU community.[16] In 2017, scientists and engineers made up 20.6% of people employed in science and technology occupations in the EU-28. Among the "technicians" subgroup, six Member States (Austria, the Czech Republic, France, Italy, Germany, and Slovakia) account for more than 50% (but less than 60%) of human resources in science and technology by occupation (HRSTO). Germany has the largest share of senior HRST with 46.8%. Besides, Germany is one of the nations that had the highest number of women employed in the science and technology profession in 2017, with 60.9%.[17]

The German school curriculum is designed to increase the proportion of college students and avoid the lack of mathematical, information technology, natural science, and technology knowledge. Due to the structure of the federal state administration (German Federal Administration), the responsibilities for research policy, research funding, scientific infrastructure, and other related activities are shared between the federal government and the 16 states. In the German innovative system structure, there is a multitude of agents interacting with each other in a complex structure on multiple levels. The STI system includes improving frame conditions relating to sponsoring R&D and innovation activities, such as providing venture capital, financial support for R&D, and innovation in SMEs. Reform takes place in the higher-education sector to ensure the quality and availability of training for young researchers and engineers, and to reduce bureaucratic obstacles for research and innovation activities. At the same time, further reforms are performed in the field of public research to strengthen technology transfer and promote links between industries and scientific research.

Research and innovation policy objectives include the rapid development of new technologies and enhanced diffusion of technologies in priority areas; the increased use and commercialization of research outcomes at research organizations; and a stronger collaboration between training institutions and the private business sector. These increase the role of SMEs in R&D and innovation activities. The stimulus is based on the formation of spin-offs.

In Germany, the Fraunhofer Associations (Fraunhofer Gesellschaft) undertakes applied research of direct utility to private and public enterprise and it is of wide benefit to society. The services of these associations are solicited from the needs of customers and contract partners in the industry, public service, and administrative sectors. The Fraunhofer Society, with 900 million euros in revenue, is the biggest applied R&D society in Germany. Sixty percent of the revenue is from contractual research, while 40% comes from public funding. In 2017, the Fraunhofer Society used 2.3 billion euros for research, which then generated 2 billion euros through contract research (Around 70% of the contract revenue is derived from contracts with industry and from publicly financed research projects, almost 30% is contributed by the German federal and states Governments)[18]

About two-thirds of the association's contract research revenue is from contracts with industry and publicly funded research projects. The remaining one-third is from federal and state government support.

Collaborative studies between the public and private sectors are widely available in various forms, such as academic funding sources (via PhD candidates and associate professors) or research projects, to transfer knowledge and technology from R&D cooperation projects. Universities receive more than 12% of the total R&D investment from the business sector. This is one of the highest figures among the 13 OECD countries. To enhance this support, most universities are actively seeking investment and involvement from the private sector. As an illustration, one can consider the case of the Technical University of Munich (TUM). Representatives of the private sector are members of the two major management organizations of TUM: the university council (Hochschulrat) and the board of directors (Kuratorium). In addition, there is a tendency to appoint experienced professors in the industry, such as in the S&T sector.

The Gross domestic expenditures on R&D of Germany in the period 2016–2018 was about 3.022% GDP.[19] According to the World Economic Forum, in 2017, a total of 128,921 patents were registered with the German Patent and Trade Mark Office (DPMA), the largest in Europe and the fifth largest in the world, and one in three patent applications in Europe came from Germany.[20]

Then, in 2018, there were four very distinct groups of countries in Europe with very different competitiveness levels and, within the EU, Germany's overall competitiveness score (82.8, 3rd) was 20 points higher than Greece (62.1, 57th).[21] Besides, Germany also maintains its ninth spot in the top 10 in the Global Innovation Index 2018, keeping its 17th position in the Innovation Input Sub-index and gaining two places in the Innovation Output Sub-Index (5th). It ranks in the top 25 economies across all pillars and in the top 10 for both output pillars. This year

Germany safeguards most of its respectable positions while improving in institutions (16th), infrastructure (19th), and business sophistication (13th). In these three pillars it improves the most in business environment (15th), ecological sustainability (31st), innovation linkages (14th), and knowledge absorption (22nd). On the output side, Germany gains only in the sub-pillar knowledge impact (17th, up four)....[22]

From 2011 until now, Germany has been a strong Innovator. In 2019, the top dimensions of Germany in innovation are intellectual assets, firm investments, and innovators. Germany performs particularly well on PCT patent applications, public-private co-publications and private co-funding of public R&D expenditures. However, human resources are the weakest innovation dimensions. Germany's lowest indicator scores are on foreign doctorate students, population with tertiary education, and venture capital expenditures.[23]

2.3   STI SYSTEM OF SOUTH KOREA

Around 1975, South Korea's per capita GDP was still comparable to the poor countries in Africa, but less than four decades later, in 2004, South Korea joined the club of trillion-dollar powerful economies. According to OECD data, during the period 2004 to 2011, the unemployment rate in the country never exceeded 3.7% while the average income per capita increased by 36% (USD30,366). With the growth strategy oriented toward developing a creative economy where innovative technology joint-venture enterprises play an important role, attracting enterprises and investors from the most advanced countries is what the Korean government has strived to do. The country had a population equal to just two-fifths of Japan's, one-seventh of the United States' and only 1/26th of the Chinese population, but Korea has multinational giant corporations like Samsung, LG, and Hyundai. These can be comparable to giant companies such as Apple, Intel, Sony, Toyota, and Ford. South Korea has had very successful high-tech models shown in innovative indicators as one of the world's leading group countries. S&T has affected all aspects of Korea's economic, social, and cultural life. The Korean STI model is indeed remarkable and worth examining for Vietnam.

In Asia, Korea is not a resource-rich country with a large workforce. In 1961, Korea was a poor country with a weak production base, a narrow domestic market, and a primarily agriculture-based economy. However, after the Korean War (1950–1953), with the demand for economic recovery and development, South Korea quickly proceeded to industrialize and modernize, creating a mechanism to promote S&T research. Korea's STI system is mainly based on a catch-up model, which prioritizes the acquisition of advanced technologies from abroad rather than investing in deep research for domestic R&D activities. According to a research of OECD (2014),[24] the Korean STI system has succeeded in accumulating research capabilities and gaining some outstanding achievements. For eight consecutive years, Korea's R&D expenditure has increased steadily. Korea's GDP accounted for 2.27% of the world's GDP and averaged USD369.07 billion from 1960 to 2014.[25] Although the world economy is recovering from the slow economic crisis in 2011, Korea still spends 4.36% of its GDP on R&D. Korea has had highest level of R&D spending among the OECD member countries.[26]

Research units of large companies (such as Samsung or Hyundai) and research institutes all receive government funding and have become the main factors that make up the Korean STI system.

According to The Global Competitiveness report 2014–2015 of *World Economic Forum*, despite having its main advantage from exports, Korea is not a mineral resource-rich country. However, with its preferential policies on finance for enterprises and innovations in education and training, it has created a momentum to continue its socioeconomic development. Korea is ranked quite high in the global competitiveness index. In the innovation criteria, Korea has a relatively high index compared to 144 countries, even higher than some developed economies. In addition, most of Korea's indicators are among the top 30 countries.[27]

The Korean economy is export-oriented. In 2012, for example, exports accounted for 57% of its GDP, which shows that Korea's level of global economic integration is quite high. In 2014, South Korea signed 10 free trade agreements (FTAs), including the ones with the United States and the EU, and is currently in negotiations with other countries. If Korea succeeds, the portfolio of its bilateral FTAs will include markets representing more than 70% of the global GDP.[28] This strategy is being applied with the participation of Korea in the Program of the 7th EU Research Framework (FP7) on research and innovation as well as in the Horizon 2020 meetings, as one of Korea's integration efforts.

The Korean innovation policy model creates an environment that requires scientists to receive intensive training and financial resources to support R&D activities that need more attention. The Korean government has launched a strategy called "Long-term Vision for S&T

Development until 2025" that outlines the directions to be implemented for building an advanced and prosperous economy through S&T development. This is done by creating, using, and disseminating knowledge, promoting scientific knowledge, and forming a progressive management system of national S&T. There are three steps in the plan:

- *Step 1 (by 2005)*: Bring S&T capabilities to the level that is competitive with the world's leading countries by mobilizing resources, expanding infrastructure, and upgrading related laws and regulations.
- *Step 2 (by 2015)*: Become an S&T development country in Asia-Pacific region, actively participating in scientific research and creating a favorable environment for promoting R&D.
- *Step 3 (by 2025)*: Ensure S&T competitiveness is equivalent to G7 countries in some areas.

These are considered to be the basic foundations for Korea's development.

In this process, the resources and results of Korea's R&D activities are mainly concentrated in a small number of industries. Among them, about three-quarters are in high-tech production and medium-high-tech industries. The effectiveness of innovation activities is reflected through the number of patents, typically five leading Korean information and communications technology companies hold 57% of the total number of Korean patents in the United States (including Samsung Electronics, which accounts for 35%). The Korean government also pays great attention to the policy of financial support for R&D human resources by supporting the payment of 80% of annual salary for each specialist, up to USD30,000 in the first two years. Many policies on tax refund and import tax reduction as well as the exemption of some taxes are reserved for high-tech scientists and enterprises. In particular, Korea does not control the results of scientific research and technological innovation based on international publications or the commercialization of products. Scientists are also able to manage the funds themselves without having to explain to the financial institution until the end of the project.

In addition, the National Research Fund of Korea provided USD770 million for building a world-class university that would be able to compete with the international community over the next five years. Funding focuses mainly on the study of strategic economic fields for international cooperation in S&T. The goal is to recruit well-known foreign scientists and researchers to work at universities. Another strategy is to promote the implementation of other forms of international cooperation such as organizing seminars, international conferences, and exchanging research work based on memoranda.

The Korean university system trains a large number of human resources, and the proportion of university-level human resources has increased rapidly. While 64% of Koreans between 25 and 34 years old have a university degree, the OECD average is only about 39%. Moreover, one-third of them graduate from universities in the fields of S&T, production, and construction. Meanwhile, this figure for the United States is 15.2%, 23.2% for Japan, 26.7% for Austria, and 30.6% for Germany.[29] It can be said that human resource is a strong basis for building and developing Korea's STI system. In addition, the Korean government has launched various initiatives to attract high-quality human resources such as the global Korean scholarship program. The number of Koreans who go abroad to study is constantly increasing, with more than 120,000 people as of 2010, while the number of foreign students who go to Korea to study was only about 6000.

In contrast to the global-oriented production and export network, Korea is placed in the lower half of the OECD ranking for co-authors and co-inventions. Korea only achieved 1.7% of patents that incorporate international research. In Asia, only Japan exceeds the OECD average with 6.8%. Meanwhile, the figure is 13.1% for Austria, 8.9% for Germany, and 6.7% for the United States.[30] Therefore, in spite of Korea's great investment in R&D, more effort is needed to improve international cooperation and knowledge-exchange indicators in an effort to internationalize the STI system.

The Korean government's strategies in international exchanges, research programs, and innovation activities in universities show the flexibility of the education system to meet the country's development needs. The effects and quality of these strategies will take a long time to show. Korea's success is due in part to its financial investment policy for STI activities. The total R&D cost of Korea in 2013 amounted to USD68,937,037 million.[31] Additionally, there are large R&D investments in the natural sciences, especially in technology.

In order to enhance R&D activities, the National Research Foundation of Korea (NRF) implemented many programs to strengthen the international research exchange between Korea and other countries. The foundation launched a sponsorship package for the "Global R&D Network", which granted USD11 million for participation in various international-ization such as the EU FP7. The Korean government also initiated a spon-sorship package, the Korean Research Fellowship, to attract world-class scientists to Korea to boost the country's economic growth and improve the quality of research at local universities. The government also intro-duced "Initiative 577" to work toward increasing total R&D spending to 5% of its GDP. In 2014, the government proposed a three-year plan of economic reform entitled "Vision 474" to maintain an annual growth rate of 4%. Accordingly, 70% of workers were employed, the per capita income was USD40,000, and GERD was raised to 5% of the GDP.[32]

On the enterprise side, large private corporations (*chaebol*) have inter-nationalized R&D activities that are more dominant than SMEs. Big companies like Samsung, LG, or Hyundai develop a network of R&D centers in different countries. Besides, the spending on R&D activities of *chaebol is* also much higher than that of SMEs.

Over the past two decades, the world has witnessed leaps in the devel-opment of Korean S&T. The right S&T policies with a high-technology orientation and the study of available research results have helped Korea grow its economy rapidly and become a bright spot of global S&T. The rapid increase in spending on R&D and innovation as well as strategies for building STI systems have helped Korea rise and hold key positions in the field of S&T. The ways in which Korea has developed its STI activities can be a good lesson for Vietnam.

2.4   STI System of Singapore

Singapore has very few natural resources and has faced fierce competition from neighboring countries for foreign investment. The Singapore gov-ernment focuses on the country's most valuable asset—knowledge—and claims S&T as the main pillar of its economy. Singapore has been open to international competition since its independence in 1965. So far, its STI achievements have led Singapore to become one of the most innovative countries in Asia.

In Singapore, schools have a major partnership with enterprises as a result of an R&D support mechanism. This is a funding mechanism aimed at stimulating research activities between schools and enterprises. In August 1992, the Industry & Technology Relations Office (INTRO) of the National University of Singapore (NUS) was set up in response to the increasing need to collaborate with industry, and to manage NUS's intellectual property as more of its work links directly with industry. INTRO's primary goal is to act as a one-stop information and service center to advice staff members on research collaboration and technology transfer. To date, INTRO has facilitated filing for more than 900 patents, over 200 of which were granted; over 150 licensing agreements were signed, bringing about USD1.2 million in revenue.[33] On average, INTRO signed about 120 annual research cooperation agreements with funding from businesses averaging under USD10 million.

2.4.1 Management System and Protection of Intellectual Property

Singapore is ranked third in the world and top in Asia for having the best IP protection in the World Economic Forum's Global Competitiveness Report 2018.[34] The Intellectual Property Office of Singapore (IPOS) under the Ministry of Law provides an environment and foundation for the creation, protection, and utilization of intellectual property products. As a member of the World Trade Organization (WTO) and, in particular, the trade section relating to intellectual property rights, Singapore has a system that protects the economic value of innovation initiatives and has very clear policies to manage intellectual property.

2.4.2 Organizational Model in Singapore's Policy Design

A characteristic feature of Singapore's STI policy design and implementation is the top-down approach. The Research, Innovation, and Enterprise Council (RIEC) counsels and advises the Singapore government on national policy as well as research and innovation strategies. It aims to make Singapore a knowledge-based economy with strong R&D capabilities that can allow the nation to promote new creative endeavors with S&T solutions and catalyze the economy's new areas of growth.[35] In 2018, The RIEC had 26 members, headed by Singapore's prime minister, 11 other members including one deputy prime minister and nine ministers, two representatives from the two leading universities, namely National University of Singapore and Nanyang Technological University. The remaining 10 members are representatives of the world's leading educational, research, and corporate organizations, such as Harvard University, the Israel Space Agency's National Research and Implementation Council, and Volkswagen Fund.

2.4.3   National Research Foundation of Singapore

The supporting organization for the RIEC is the National Research Foundation (NRF), whose duties include assisting the former; coordinating research activities of organizations and institutions nationwide in a clear overall strategic direction; developing policies and plans to implement the five incentives for national research and development programs; implementing research, innovation, and business strategies approved by the RIEC; and allocating funds to programs that meet the NRF's strategic goals. The NRF has close ties with ministries, especially the Ministry of Industry and Trade, Ministry of Health, Ministry of Education, and the Ministry of Environment and Water Resources. It is also closely linked with research funding institutions at lower management levels, including the Economic Development Board, the Agency for Science, Technology, and Research (A*STAR), the Academic Research Council, and the National Health Research Council. The NRF is also the focal point that provides research funding for institutes, universities, hospitals, public and private laboratories, industries, and businesses.

NRF Executive Board: The NRF executive board comprises 24 members, 17 of whom are leaders of government organizations, with one deputy prime minister and seven ministers. Of the remaining seven members, four are chancellors of top Singapore universities, and three are Singaporean industry representatives.

The Science Advisory Board is tasked to clarify essential issues, and, in collaboration with the NRF, identifies new global trends in basic science that Singapore needs and is capable of researching on. This board gives support and advice on R&D management to the NRF, including funding allocation and evaluation of research results.

It can be seen that, with this top-down approach in management, Singapore's STI policies are built with close links and are part of a broader economic development strategy. Up until now, Singapore does not have a separate ministry of science and technology, but instead the planning and implementation of STI policies are included in the economic development policy.

Orientated toward prioritizing investment for STI, Singapore's system of policies to support S&T is quite comprehensive and methodical. The Singapore government has implemented many policies to support commercialization, business development, tax incentives, and risk insurance. Some of Singapore's key STI policies on scientific research, human resources, and foreign investment are described in section 4.4.

2.4.4   Policy to Increase Investment in Scientific Research

Policy makers say that a nation's success is due to its bidding policy, in accordance with international standards for scientific research. In fact, the Singapore government has not created any new models, but it mainly relies on existing systems, especially the British model. Accordingly, the government implements a bidding policy for projects of three, five, or 10 years, from the point of view of technology transfer through cooperation with industrialists and the purchase of patents to further accelerate the process from discovery to application.

Meanwhile, in order to compete effectively with emerging markets, Singapore has focused its R&D capabilities in both the public and private sectors, through the National Science and Technology Plan 2010. This plan was launched in 2006, which saw the investment of SGD12 billion to support the development of the plan, thereby increasing R&D spending to 3% of the GDP.[36]

Singapore has taken a sustained approach to funding R&D as a critical pillar of Singapore's economic development strategy. In 2014, Singapore's annual public expenditure on R&D was SGD$3.3 billion, a compound annual growth rate of 11.1% over the past nearly two and a half decades (1990–2014). Over the same period, the annual business expenditure on R&D has grown at a compound annual growth rate of 12.5%.[37] The country's public expenditure on R&D accounted for 0.8% of GDP, while the total business expenditure on R&D of Singapore amounted to $5.2 billion, corresponding to 1.3% of Singapore's GDP in 2014.[38]

Experimental development accounted for 51% of the total R&D expenditure, 30% of which was applied research and 19% was basic research. Engineering and Technology received 62% of the total R&D spending, 11% for Natural Sciences (excluding biological sciences), 18% for Biomedical and relevant sciences, 1% in the Agricultural and Food Sciences, and 8% in other areas.

2.4.5   Policies to Attract High-Quality Human Resources

Singapore has promoted an open policy to attract specialized talents from around the world. Foreign workers in Singapore account for a large number of the total workforce. This has been demonstrated by the Worldwide Quality of Living Index 2007 by Mercer Human Resource Consulting, which ranked Singapore at the top of Asian countries in terms of quality of life. At the same time, the 2007 ECA International survey also showed that Singapore is the world's best city for Asians who want to live, work, and enjoy life abroad. The strong management of the government has created a stable environment in which to work. In 2013, the World Bank ranked Singapore as the best country to work in the world. According to the *APO Productivity Databook 2012* on the productivity of Asian countries, in 2010, Singapore was the country with the highest labor productivity, reaching USD89,900/worker (calculated by purchasing power parity in 2005).[39] On the 2015 global creative ranking, Singapore ranked ninth, and in the labor efficiency index ranked fifth with an index of 0.889.

Yet, Singapore is still facing the challenges emerging from the lack of experts in many important areas. Those experts who are capable of handling complex tasks need to have multidisciplinary knowledge, not only technical knowledge, legal knowledge, but also a proper understanding of social movement.

2.4.6   Policy to Attract Foreign Investment

Singapore's strategy focuses on attracting foreign investment, which provides an opportunity to improve its technological capacity and innovation. Singapore's rapid economic growth and extended scaling have been achieved by continued restructuring and improvement.[40]

According to *Doing Business 2009* (the World Bank's annual report that compares management regulations of 181 countries), Singapore was ranked first for having the most favorable business environment, and held this position for three consecutive years.[41] The World Economic Forum report on Global Competitiveness 2017–2018 acknowledged Singapore as the most competitive economy in Asia and third in the world.[42] Businesses around the world see Singapore as the ideal place to grow their business, many of whom consider Singapore to be a stepping stone to enter emerging Asian markets.

The successful operation of Singapore's STI system is due to the right development orientation, with a national development strategy that has an extensive vision, appreciating the capacity of the country, focusing on human resources education and training, which are considered key issues, while taking advantage of external resources in the development process. In that process, Singapore learned and benefited from the best experiences of the most advanced countries in the world, and applied those experiences to the local context.

In short, globalization and the knowledge economy are turning the world into a single market emphasizing talent and intelligence. But not every country has the ability to provide an environment full of opportunities and an international environment like Singapore. Since independence, Singapore has overcome many challenges to become one of the most developed economies in Asia. Many entrepreneurs move to Singapore to live and work, while there are still countless opportunities open to global citizens who are able to apply their skills and knowledge. Singapore has achieved high-quality trained human resources and a sustainable growth economy based on STI development, and this will continue to be the direction of Singapore in the future.

2.5 Lessons for International Integration Process of Vietnam's STI Policy

International integration has been an indispensable trend contributing to the development of all countries. Countries with different economic, political, and social contexts have their own ways of choosing international integration, receiving opportunities, and facing the challenges of this process. For Vietnam, the process of international integration, especially the international integration of STI, is slower than some countries in Asia and the world. Apart from the progress achieved from the process of international S&T integration, Vietnam also needs to make adjustments in its strategic development direction.

2.5.1 Lessons Learned from STI Systems of Other Countries for International S&T Integration in Vietnam

Increasing globalization and competition between countries require a country to make appropriate changes to adapt to the new situation and enhance national competitiveness through technology development.

First of all, attention should be paid to the role of the state in the STI system. No country in the world could thrive in STI without the active participation of the government. For Vietnam, the role of the government in the STI system should be defined as a macro-management role by policies. The state manages and promotes science through development policies in an innovative and modern manner. The state plays the role of building a legal framework for science, orienting key areas and priorities, making short- and long-term development plans. At the same time, the state has policies to prioritize development with key areas as well as separate policies for science, especially financial policies for S&T to build autonomy in science. In the current period especially, it is necessary to focus on developing applied sciences to create highly competitive technology products in the market.

In terms of system resources, finance plays an important role in the survival of the STI system. Therefore, the state needs to develop a reasonable spending policy to promote S&T development and reform the management method of budget expenditure to promote S&T development, with the following solutions:

- Strengthen autonomy for units and beneficiaries of investment for S&T.
- Promote the process of managing state budget spending for S&T according to the funding model. With the funding model, financial resources for S&T are not divided every year. Money is provided in a timely manner to carry out tasks and promptly solve urgent requirements set by the socioeconomic development practice. At the same time, administrative procedures for capital allocation and settlement for S&T activities will be simplified.
- Need to increase investment in R&D activities of universities and research institutes. This measure is likely to contribute greatly to productivity growth, which is reflected in:

  - Creating a by-product of the research process, which is a team of S&T scientists trained in the research process.
  - Results of the research process are spread through the relationships of schools and institutes with enterprises.
  - Universities, research institutes, and laboratories are the idea incubators, creating new businesses and enterprises because they are the first to observe the potential of commercializing research results.

- Innovate the inspection and supervision of the use of the state budget for S&T fields. In order to check and evaluate the efficiency of capital use, attention should be paid to evaluating the outputs that

S&T activities have achieved. Eliminate assessment based on compliance with spending norms and time frames.

- Mobilize social resources to participate in developing the national STI system. Research funding through venture funds has been a successful experience that Vietnam can learn from.

Next is the creation of close and sustainable links between universities, enterprises, and research institutes. The experience of the countries studied is to take advantage of innovative ideas and research from universities, research institutes, and leading experts in the fields, and link them to market requirements. Enterprises capture market demands to orient research activities, as well as creative ideas from universities and research institutes to create new breakthrough demands for the market. The academic–industrial relationship is strongly promoted through the investment efficiency for STI, which is significantly heightened.

Vietnam also needs to focus on training S&T human resources in universities, especially in high-quality human resources that can meet the increasing requirements of qualifications to reach the intellectual level of advanced countries. In addition, it is essential to build an adjacent team in S&T research because human resources are the national competitive resources. In the policy of national S&T human resources, Vietnam is advised to have a stronger policy to attract more high-quality human resources to Vietnam and to facilitate research and work to retain talents and prevent an outflow of talent.

In recent years, Vietnam's STI has undergone strong development. It has taken advantage of international integration processes such as technology transfer and attracted high-quality human resources from other countries to study and work. It has also strengthened S&T cooperation relations to create a momentum for development and strengthened internal capacity for domestic S&T, and applied the development experiences in other countries. It is essential to have an appropriate application of the countries' development experiences to Vietnam's situation in order to develop Vietnam's STI in the face of challenges and increasingly fierce competition.

2.5.2 Trends and Impacts of the Development of the World's STI in Vietnam

In recent years, the economic and political situation in the world has entailed many changes, impacting almost all fields. In addition, the knowledge economy, which is characterized as an economy where knowledge plays a key role (OECD 1996), has been prioritized in the nations' development orientation. In particular, economic development is related to the level of technological competition, which depends on science and scientific research. Vietnam is in the process of integrating deeply into the world economy; above all, globalization, as an inevitable trend, is a strong influence on all countries. For STI activities especially, Vietnam is influenced by the world's development trends.

2.5.2.1 Formation of Knowledge Economy
The wave of innovation is and will be happening in a series of S&T fields such as software technology, computers, telecommunications, and artificial intelligence.

The knowledge economy is an economy where growth and development no longer rely primarily on natural resources as before, but are mainly based on S&T knowledge sources (types of resources capable of reproduction and self-reproduction, and which is never exhausted). The knowledge economy has characteristics that directly affect the choice of the S&T development orientation of a country. They are:

- High content of knowledge and information in products and services.
- Development based on knowledge and information.
- Network connectivity and integration in economic activities.
- Human factor, the role of knowledge, and skills of human resources are decisive factors in competition and development.
- Flexibility in economic adjustment and development.

For developing countries like Vietnam, the trend toward a knowledge-based economy can offer the following advantages to the country's S&T development:

- The training of human resources in new and specialized S&T fields of the knowledge economy, such as biotechnology and information technology, which are Vietnam's strengths.
- Select and absorb the latest S&T achievements toward the knowledge economy to exploit and promote the country's comparative advantages in relation to regional and world economies.

2.5.2.2 Enhanced Links Between STI System Components

There should be a link between research institutes, universities, and businesses. This three-party association will create an understanding of mutual mechanisms and needs, on the basis of "ordering" one another. It is a model for building STI systems and is currently being implemented by many countries worldwide. The linkage of schools, institutes, and businesses is a global trend positively affecting the development of the parties as well as the general development of national S&T. In order to catch up with developed countries, it is necessary to further promote the links between schools, institutes, and enterprises into networks of research, training, and production of technology products, especially high technology.

2.5.2.3 Promoting Financial Investment for STI

When S&T activities are taken into account, financial resources are often emphasized. In the context of global integration, innovation activities produce many advantages. In line with development requirements, R&D investment in countries around the world has tended to increase.

Nations' innovation policies have always aimed to encourage R&D investment in the business sector, where innovation is the most powerful. Through businesses, research organizations, and many international organizations, R&D investment from foreign sources tends to increase. When the world forms a value chain due to international economic integration, many multinational companies would move R&D activities to the countries in which they invest.

The trend of strong transfer of foreign investment and strong shift of R&D activities of multinational companies to foreign enterprises and branches has contributed significantly to improving the value of manufacturing and exporting high-tech products in developing economies.

2.5.2.4 Toward Sustainable Development

Sustainability in development today becomes the leading demand and requirement of every country in the world. The concept implies socioeconomic and ecological environment sustainability. The concept of sustainable development has been popular since 1987, stemming from the World Commission on Environment and Development's report. Sustainability is seen as "a human goal of human-ecosystem equilibrium", and sustainable development involves a "holistic approach and processes that temporarily bring people to the end point of sustainability".[43]

For Vietnam to emerge as a modern industrialized country and achieve the average per capita income in the group of countries with average development level in the world by 2020, are those strategic goals feasible, and what will S&T activities do to contribute practically to the implementation of those strategic directions? It can be said that these are difficult and complex tasks whose implementation requires change in S&T management and implementation. For Vietnam, in addition to examining development trends in the world to take advantage of external resources, attention should be paid to the forecasting of S&T development in order to develop STI development strategies at the national level that are suitable for the local context. Attention should also be paid to prioritizing investment and demonstrating the progress and conformity with the general trend of the world.

Some conclusions can be drawn from researching and analyzing the experiences of selected countries: (1) institutional reforms facilitate the development of S&T; (2) it is important to implement key policy measures such as training human resources and attracting S&T talents (e.g. Sweden and Singapore), issuing financial and credit policies to facilitate technological innovation activities of the enterprise sector, and making enterprises the main subject of technological innovation; (3) investment in S&T needs to be increased; and (4) there needs to be focus on international cooperation in S&T.

In general, all the four conclusions mentioned above are recommended for Vietnam, particularly in terms of institutional reforms for S&T development, as the "joints" between S&T activities and the application of S&T results created in production and business are still loose. Enterprises are still weak in the potential and motivation for technological innovation (Dao Thanh Truong 2016). Meanwhile, a number of S&T organizations are confused about the process of transforming into S&T enterprises or operating under the enterprise mechanism. Moreover, businesses, research institutes, and universities have not yet been allied in any association or coordination that could otherwise create value in the STI system. In addition, Vietnam needs to quickly build and train human resources to serve the operation of the STI system, focusing more on mobilizing investment capital for S&T activities and gradually forming the philosophy of scientific management.

NOTES

1. https://www.worldometers.info/world-population/sweden-population/ (19/06/2019).
2. https://ssci.se/en/news/swedish-innovation-system-unique (19/06/2019).

3. Research and Innovation Bill of Government of Sweden, pp. 14–15.
4. Triple helix is a term used when discussing innovation and creativity policies with three main participating forces: entrepreneurs (the business community or owners of innovations); research institutes or universities (organizations creating knowledge, supporting the process of innovation), and the state (which supports the process of innovation through funding or technical assistance). These three forces create a tripodal posture where they need to work closely together to strengthen the innovation system. The model of the triple helix three-legged tripod is considered to have been successfully applied in the Netherlands and Finland.
5. Stefan Einarsson and Filip Wijkström (2015). European Foundations for Research and Innovation EUFORI Study. Publications Office of the European Union.
6. Roger Svensson—Mälardalen University & IFN—Research Institute of Industrial Economics (2008). Vinova report (2008). Growth through research and development—what does the research literature say?
7. https://en.wikipedia.org/wiki/Bank_of_Sweden_Tercentenary_Foundation.
8. OECD (2016). *OECD Reviews of Innovation Policy: Sweden 2016*, p. 65.
9. Eurostat (2019). First estimates of Research & Development expenditure. R&D expenditure in the EU increased slightly to 2.07% of GDP in 2017. Two-thirds was spent in the business enterprise sector. New release from 5/2019–10 January 2019.
10. https://www.wipo.int/pressroom/en/articles/2018/article_0005.html.
11. Read more at https://knowledge.insead.edu/entrepreneurship/the-worlds-most-innovative-countries-2019-12016#szAHAy76JOBY4A3Y.99.
12. https://sweden.se/business/innovation-in-sweden/.
13. Government Office of Sweden (2012). The Swedish Innovation Strategy 2020, p. 21.
14. EU Scoreboard Innovation Index 2015, p. 10.
15. Global Innovation Index 2018, Germany. Read more in: https://www.wipo.int/edocs/pubdocs/en/wipo_pub_gii_2018-profile15.pdf.
16. OECD (2013), *OECD Science, Technology and Industry Scoreboard 2013*.
17. https://ec.europa.eu/eurostat/statistics-explained/index.php?title=Human_resources_in_science_and_technology_-_stocks&oldid=76959#Professionals_and_technicians_employed_in_science_and_technology_occupations.
18. Marianne Hoffmann (2018) presentation on the Fraunhofer model technology transfer from universities to industry in Japanese-German symposium 26–27.4.2018, Tokyo.
19. https://data.oecd.org/rd/gross-domestic-spending-on-r-d.htm.

20. https://www.weforum.org/agenda/2018/10/germany-is-the-worlds-most-innovative-economy/.
21. Klaus Schwab (Editor) World Economic Forum (2018). *The Global Competitiveness Report 2018*.
22. Soumitra Dutta, Rafael Escalona Reynoso, Antanina Garanasvili, and Kritika Saxena, SC Johnson College of Business, Cornell University Bruno Lanvin, INSEAD Sacha Wunsch-Vincent, Lorena Rivera León, and Francesca Guadagno∗, WIPO (2018). *The Global Innovation Index 2018: Energizing the World with Innovation*, p. 17.
23. European Innovation Scoreboard 2019, Germany, p. 47.
24. OECD Economic Outlook, No. 95 (2014), *OECD Economic Surveys KOREA Overview*.
25. OECD (2014), *OECD Science, Technology and Industry Outlook 2014*.
26. Olbrich, P., & Witjes, N. (2014). Rethinking Borders of National Systems of Innovation: Austrian Perspectives on Korea's Internationalization of Green Technologies. *STI Policy Review*, 5(2), 65–95.
27. Klaus Schwab (2014), *The Global Competitiveness Report 2014–2015*, World Economic Forum.
28. OECD (2014), *OECD Science, Technology and Industry Outlook 2014*.
29. Olbrich, P., & Witjes, N. (2014). Rethinking Borders of National Systems of Innovation: Austrian Perspectives on Korea's Internationalization of Green Technologies. *STI Policy Review*, 5(2), 65–95.
30. OECD (2014), *OECD Science, Technology and Industry Outlook 2014*.
31. OECD (2014), *OECD Science, Technology and Industry Outlook 2014*.
32. Anthony Fensom (2014), *Park Warns South Korea: Change or Perish*, http://thediplomat.com/2014/02/park-warns-south-korea-change-or-perish/.
33. Jasmine Kway (2013), University and Industry Relations in Singapore. The translated document of Center for Higher Education Evaluation and Research, Nguyen Tat Thanh University, Vietnam. International Education Information Magazine, No. 10, 2013.
34. See IPOS (2019), Singapore's IP ranking.
35. OECD (2011), *Review of Innovation in South-East Asia Country Profile: Singapore*.
36. Ministry of Trade and Industry Singapore (2006), *Science and Technology Plan 2010*, SNP Security Printing Pte Ltd.
37. Cornell University (2016), The Global Innovation Index 2016: Winning the Global Innovation, WIPO.
38. Agency for Science, Technology and Research Singapore (2015), National Survey Of R&D In Singapore 2014, Agency for Science, Technology and Research Singapore.
39. Asian Productivity Organization (2012), *APO Productivity Data Book 2012*, Keio University Press Inc, Tokyo.

40. Sanjaya Lall, Shujiro Urata (2003), *Competitiveness, FDI and Technological Activity in East Asia*, Edward Elgar Pub.
41. The World Bank (2008). *Doing Business 2009*, Washington, DC: The World Bank.
42. Klaus Schwab (2017), *The Global Competitiveness Report 2017–2018*, World Economic Forum.
43. Shaker, Richard Ross (September 2015), *The Spatial Distribution of Development in Europe and Its Underlying Sustainability Correlations*, Applied Geography. 63. p. 35. https://doi.org/10.1016/j.apgeog.2015.07.009.

REFERENCES

Agency for Science, Technology and Research Singapore. (2015). *National Survey of R&D In Singapore 2014*. Agency for Science, Technology and Research Singapore.

Asian Productivity Organization. (2012). *APO Productivity Data Book 2012*. Tokyo: Keio University Press Inc.

Cornell University. (2016). *The Global Innovation Index 2016: Winning the Global Innovation*. WIPO.

Dao Thanh Truong. (2016). *Chính sách Khoa học, Công nghệ, và Đổi mới (STI) của Việt Nam trong xu thế hội nhập quốc tế: Thực trạng và giải pháp* [Science, Technology, and Innovation Policies of Vietnam in the Trend of International Integration: Situations and Solutions]. Hanoi: Thế Giới Publishers.

Dutta, S., Reynoso, R., Garanasvili, A., Saxena, K., Lanvin, B., Wunsch-Vincent, S., León, L., and Guadagno, F. (2018). The Global Innovation Index 2018: Energizing the World with Innovation. In: S. Dutta, B. Lanvin, and S. Wunsch-Vincent, ed., *The Global Innovation Index 2018: Energizing the World with Innovation*, 11th ed. Ithaca, Fontainebleau, and Geneva: Cornell University, INSEAD, and WIPO, p. 17.

Einarsson, S., and Wijkström, F. (2015). *European Foundations for Research and Innovation EUFORI Study*. Publications Office of the European Union.

Eurostat. (2019). *First Estimates of Research & Development Expenditure*. Eurostat Press Office. Available at: https://ec.europa.eu/eurostat/documents/2995521/9483597/9-10012019-AP-EN.pdf.

Fensom, A. (2014). *Park Warns South Korea: Change or Perish*. [Online] The Diplomat. Available at: http://thediplomat.com/2014/02/park-warns-south-korea-change-or-perish.

Government Office of Sweden. (2012). *The Swedish Innovation Strategy 2020*, p. 21.

Jasmine Kway. (2013). *University and Industry Relations in Singapore*. Translated by Center for Higher Education Evaluation and Research, Nguyen Tat Thanh

University, Vietnam. *International Education Information Magazine*, No. 10, 2013.

Lall, S., and Urata, S. (2003). *Competitiveness, FDI and Technological Activity in East Asia*. Cheltenham: Elgar.

Marianne Hoffmann. (2018). Presentation on the Fraunhofer Model Technology Transfer from Universities to Industry in Japanese-German Symposium 26–27.4.2018, Tokyo.

Ministry of Trade and Industry Singapore (2006). *Science and Technology Plan 2010*. SNP Security Printing Pte Ltd.

OECD. (1996). *The Knowledge-Based Economy*. Paris: OECD.

OECD. (2011). *Review of Innovation in South-East Asia Country Profile: Singapore*. Paris: OECD.

OECD. (2013). *OECD Science, Technology and Industry Scoreboard 2013*. Paris: OECD.

OECD. (2014). *Overview, OECD Economic Surveys KOREA*. Economic Outlook, No 95.

OECD. (2016). *OECD Reviews of Innovation Policy: Sweden 2016*, OECD, p. 65.

Olbrich, P., and Witjes, N. (2014). Rethinking Borders of National Systems of Innovation: Austrian Perspectives on Korea's Internationalization of Green Technologies. *STI Policy Review*, 5(2), pp. 65–95.

Roger Svensson—Mälardalen University & IFN—Research Institute of Industrial Economics. (2008). Vinova Report (2008). Growth Through Research And Development—What Does the Research Literature Say?

Schwab, K. (2014). *The Global Competitiveness Report 2014–2015*. World Economic Forum.

Schwab, K. (2017). *The Global Competitiveness Report 2017–2018*. World Economic Forum.

Schwab, K. (Editor) (2018). *The Global Competitiveness Report 2018*. World Economic Forum.

Shaker, R. (2015). The Spatial Distribution of Development in Europe and its Underlying Sustainability Correlations. *Applied Geography*, 63, pp. 304–314. https://doi.org/10.1016/j.apgeog.2015.07.009.

The Intellectual Property Office of Singapore. (2019). Singapore's IP Ranking. Available at: https://www.ipos.gov.sg/who-we-are/singapore-ip-ranking.

The World Bank. (2008). *Doing Business 2009*. Washington, DC: The World Bank. Available at: https://data.oecd.org/rd/gross-domestic-spending-on-r-d.htm

https://ec.europa.eu/eurostat/statistics-explained/index.php?title=Human_resources_in_science_and_technology_-_stocks&oldid=76959#Professionals_and_technicians_employed_in_science_and_technology_occupations

https://knowledge.insead.edu/entrepreneurship/the-worlds-most-innovative-countries-2019-12016#szAHAy76JOBY4A3Y.99

https://sweden.se/business/innovation-in-sweden/

https://www.weforum.org/agenda/2018/10/germany-is-the-worlds-most-innovative-economy/
https://www.wipo.int/edocs/pubdocs/en/wipo_pub_gii_2018-profile15.pdf
https://www.wipo.int/pressroom/en/articles/2018/article_0005.html
https://www.worldometers.info/world-population/sweden-population/ (19/06/2019)

CHAPTER 3

Vietnam's STI System in the International S&T Integration Context

3.1 VIETNAM'S STI SYSTEM IN INTERNATIONAL S&T INTEGRATION

3.1.1 Characteristics of Vietnam's STI System

3.1.1.1 Structure

Presently, there are many different approaches for analyzing innovation systems such as the NIS, regional innovation system, and industry innovation system. According to a study by the Kraemer-Mbula and Wamae (2010), the innovation system consists of several key elements such as organizations, institutions, and interactive links.[1] Of these, businesses and other organizations such as universities, venture capital funds as well as policy enforcement and planning agencies, play a significant role in the NIS.

Vietnam's NIS also includes the above elements, where the main set of organizations includes: (1) *central management agencies system*; (2) *assisting ministries, departments, and advisory bodies*; (3) *S&T organizations (schools and research institutes)*; (4) *S&T enterprises*, and (5) *investment funds for innovation activities*.

According to research by the OECD (2014), the process of implementing STI activities in Vietnam has gone through many stages. Since 2011, Vietnam has been moving toward a full and complete innovation system, where it is expected to promote the role of enterprises to become a core

© The Author(s) 2019   77
D. T. Truong, *Perspectives on Vietnam's Science, Technology, and
Innovation Policies*,
https://doi.org/10.1007/978-981-15-0571-3_3

component of national innovation activities. Up until now, however, the contribution of S&T public institutions at research institutes and universities is still dominant in Vietnam's NIS, which has received great support from the state. The pervasiveness of innovation activities in S&T public institutions is modest because most S&T organizations have not yet converted completely to autonomous and self-responsible mechanisms, but still use the state budget. The development philosophy of NIS remains the 'S&T push policy', with the state being dominant in all activities instead of shifting to the market-pull philosophy in accordance with market regulations. Even so, Vietnam's NIS is fully involved with the organizational components as in the innovation systems of developing countries.

1. *System of central agencies*

The Communist Party of Vietnam, the National Assembly, and the government play an important role in determining the development paths of the country, where S&T development is considered *the top national policy, the most important driving force for developing modern production forces, knowledge economy, improving productivity, quality, efficiency and competitiveness of the economy; environmental protection, ensuring national defense and security.*[2]

Central agencies have an important role in creating institutions to implement strategies for STI development to serve the country's socioeconomic development objectives. The important ideas of S&T development and building a national renovation system were mentioned in the National Assembly's Resolution No. 142/2016/QH13, dated 12 April 2016, in the five-year socioeconomic development plan for 2016 to 2020 and in the government guidelines for national development. One of the important milestones in promoting the transformation of S&T management is the introduction of Decree 115/ND-CP to form autonomous institutions in S&T activities. At present, however, the philosophy of 'the state doing science' is still maintained. This is a difference with the national innovation systems in other countries, where the state only performs the role of facilitating the development of components of the NIS, with a focus on businesses.

2. *Ministries, departments, assisting agencies, and advisory councils*

The ministries, departments, and assisting agencies play a role in implementing the Party's policy as well as legal documents of the state and the government on S&T development.

The Ministry of Science and Technology of Vietnam performs the function of S&T state management. Its tasks include scientific research; technology development; innovation activities; developing S&T potential; managing intellectual property; enhancing quality measurement standards; managing atomic energy, radiation, and nuclear safety; and management of public services in the fields managed by the ministry in accordance with law.[3] While the policies promulgated mainly consider innovation as the target of implementation, Vietnam has not yet had innovation policies that are clearly separated from S&T development policies. The policies to promote technology development have also not yet been integrated into industrial, investment, and service development policies (Hoang Van Tuyen, 2017).[4] In recent years, for the purpose of improving the efficiency of deployment activities of innovation, the National Ministry of Science and Technology has been actively negotiating, signing, and implementing international treaties and agreements with countries with strong S&T potential. It has also been deploying Official Development Aid projects in S&T fields and innovations,[5] contributing to the mobilization of foreign investment resources. This has made a positive impact on the NIS, innovation, and entrepreneurship ecosystem in Vietnam.

The National Council for Science and Technology Policy is the advisory body of the prime minister for Vietnam's policies on S&T development in a socialist-oriented market economy. The council advises the prime minister on two levels: independently, and on demand.[6]

3. S&T organizations

Vietnam has a system of S&T organizations covering a wide range of S&T activities.

According to the Law on S&T 2013 (Article 9), the form of an S&T organization is stipulated as follows:

(a) Scientific research organizations and scientific research and technological development organizations include academies, institutes, centers, laboratories, research stations, observation stations, testing stations, and other forms stipulated by the Minister for Science and Technology.

(b) Higher-education institutions are organized according to the provisions of the Higher Education Law.

(c) S&T service organizations take the form of centers, offices, laboratories, and others prescribed by the Minister for Science and Technology.

Also, under the Law on S&T 2013, S&T organization is classified as follows:

- *By the authority of establishment, S&T organizations include*:

    - S&T organizations under the National Assembly and the National Assembly Standing Committee (established by the National Assembly and the National Assembly Standing Committee).
    - Government S&T organization (established by the government).
    - S&T organization of the Supreme People's Court (established by the Supreme People's Court).
    - S&T organization under the Supreme People's Procuracy (established by the Supreme People's Procuracy).
    - S&T organizations under ministries, ministerial-level agencies, and governmental agencies (established or authorized by the prime minister to ministers, heads of ministerial-level agencies, and governmental agencies).
    - S&T organizations under ministries, ministerial-level agencies, and governmental agencies (established by ministers, heads of ministerial-level agencies, and governmental agencies, except for cases described above).
    - Local S&T organizations according to their competence (established by provincial-level People's Committees).
    - S&T organizations established by political organizations, sociopolitical organizations, social organizations, and socio-professional organizations according to law provisions and charters.
    - Other enterprises, organizations, and individuals setting up their own S&T organizations.

- *By functions*, S&T organizations include basic research organizations, applied research organizations, and S&T service organizations.
- *By the form of ownership*, S&T organizations include public S&T organizations, non-public S&T organizations, and S&T organizations with foreign capital.

According to statistics of the Ministry of Science and Technology, in 2014, Vietnam had 1055 S&T organizations. The number of R&D organizations accounted for 47.9%, followed by tertiary education institutions comprising 32%, and finally S&T service organizations accounting for 20.1%. In addition to the system of public S&T organizations concentrated in government agencies and ministries, there were local S&T organizations. As of 2015, 310 S&T organizations were under the management of the Department of Science and Technology. The total number of employees was 7066, and the total expenditure for regular tasks was VND1243 billion (approximately USD54 million).[7]

(a) *Public research and deployment organizations*

Vietnam's system of public research and development organizations include:

  - Government academies of science.
  - R&D organizations established by the government, the prime minister, ministers, ministerial-level organizations, and governmental organizations (excluding two academies).
  - R&D organizations under the provincial People's Committee.
  - R&D organizations established by state organizations and public non-business units according to their competence.
  - R&D organizations under state-owned groups and corporations.

According to the statistics by the Ministry of Science and Technology of Vietnam, R&D organizations mainly operate in the fields of S&T (35%), most of which are located in Hanoi and Ho Chi Minh City (nearly 70%). The number of large-scale S&T public organizations (over 300) is only about 3%.[8]

In addition, some types of R&D organizations are prioritized by the state to invest in projects such as key laboratories and incubators. As of 2015, the country had 16 laboratories that have been under construction and put to use,[9] and more than 40 technology incubators and S&T enterprises, mainly concentrated in Hanoi and Ho Chi Minh City. By 2016, the number of incubators tended to increase and concentrate in high-tech zones. In the Hoa Lac Hi-Tech Park, its High-tech Enterprise Incubation Center had 37 groups registering for incubators, 10 of which were postgraduate student groups, four post-incubation groups, and 17 preincubation groups. In the Business Incubator of the Ho Chi Minh City High-Tech Zone, by 2016, 12 projects had been incubated, raising the total number of projects that are incubating to 18. The total revenue of incubating enterprises reached more than VND16 billion.

(b) *Universities, academies, and colleges*

Higher-education institutions are among the key components of the STI system. Universities and colleges can form innovative clusters in the S&T community, and act as a bridge between businesses, research institutes, and the state, or among countries. This component performs the function of developing and training young scientists, providing them with specific skills and knowledge that can contribute to the economy and create interest in scientific research and technology innovation. Above all, this is a cradle to train young human resources with skills and qualifications for businesses to improve competitiveness and flexibility in the global knowledge economy and international S&T integration. The university plays an important role in the creation of intellectual property including inventions/research results and thereby contributes to the promotion of innovation. The Innovation Partnership Program's view is that the key to promoting innovation is to increase circulation and association, and coordinate implementation among actors in society, mainly to facilitate coordination and association in innovation among universities/research institutes with businesses and the state.

In Vietnam, universities (regional and national), institutes, and colleges (hereinafter referred to as universities) are responsible for conducting R&D and combining training with scientific research and production as well as S&T service under the provisions of the Higher Education Law (2012), the Law on S&T (2013), and other laws. The university performs basic research tasks, key S&T tasks of the state, and scientific research on education.

In addition, a strategic academy model of STI and some university models linking research and production (private universities owned by large corporations) have emerged in Vietnam.

In 2018, the Vietnam Institute of Science, Technology and Innovation was established to reorganize and rearrange the National Institute for Science and Technology Policy and Strategy Studies (NISTPASS) and the Management Training Institute (MTI).

Box 3.1 Model of Vietnam Institute of Science, Technology and Innovation

The Vietnam Institute of Science, Technology and Innovation is a public scientific research institute under the Ministry of Science and Technology, which performs the function of STI research; training for master's and doctoral degrees; fostering professional management of STI; supporting and implementing innovative and entrepreneurial innovation activities; consulting; and providing S&T services. The institute consists of non-business units (including research institutes, training centers, S&T service centers) and autonomous units (including S&T enterprises, nurseries, and laboratories).

Source: https://www.most.gov.vn/en/news/123/Functions-and-tasks.aspx

In addition to the development of innovative activities in public schools, large corporations and businesses are now aware of the role of building a team of high-quality human resources, with qualifications, skills, and the professional service to achieve business goals in innovation and international integration.

Box 3.2 Model of FPT University

FPT University was established on 8 September 2006 under Decision No. 208/2006/QD-TTg by the prime minister and is operating under the Regulation on Organization and Operation of Private Universities under Decision No. 61/2009/QD-TTg issued on 17 April 2009 by the government. FPT University belongs to FPT Group, one of the largest information technology (IT)-telecom groups in Vietnam's private sector. The difference between FPT University and other universities is the form of training it has in close connection with enterprises, linking training with implementation, and performing R&D activities with modern technology. FPT's human resources are trained to focus on knowledge, skills, and foreign languages for international integration. The immediate goal of FPT University is to train and provide high-quality human resources specialized in IT, economics, fine arts, and other industries for domestic businesses, as well as world corporations.

Source:http://hanoi.fpt.edu.vn/gioi-thieu/lich-su-thanh-lap-dai-hoc-fpt

(c) *S&T service organizations*

S&T service organizations are organized in the form of centers, offices, laboratories, and others, with the main function of conducting service activities and technical support for R&D activities; activities related to industrial property and technology transfer, standards, technical regulations, measurement, product quality, goods, nuclear radiation safety, atomic energy; and services of information, consultancy, training, retraining, dissemination, and application of S&T achievements in socioeconomic fields.

S&T service organizations mainly focus on the S&T field (more than 60%), followed by social sciences and humanities (16.5%), natural sciences (12.3%), agricultural science (5.7%), and finally medical and pharmaceutical science (1.4%). Based on geographical distribution, S&T service organizations are distributed widely throughout Vietnam, with the highest concentration in Hanoi and the Red River Delta region (35.3%), followed by the North Central Coast and Central Coast (18.4%), Ho Chi Minh City and the Southeast (19.3%), the Mekong River Delta (11.8%), the Northern Midlands and Mountains (10.4%), and the lowest is the Central Highlands (1.9%).[10]

4. *S&T enterprises*

As of 1 July 2018, the country had 702,710 existing businesses. As of 31 December 2017, the state-owned enterprises sector (including enterprises with 100% state capital, and enterprises with state capital accounting for over 50%) had 2486 enterprises operating, decreasing by 6.6% compared to 2016. Non-state enterprises had 541,753 enterprises, accounting for 10.9%, and the FDI sector had 16,178 enterprises, comprising 15.5% (General Statistics Office, 2018).

Vietnam has not yet formed a complete market economy, and state-owned enterprises have not suffered much competitive pressure, so the pressure of innovation is not great. While FDI enterprises choose to hire R&D experts or transfer technology from more developed countries, the spread of R&D in Vietnam is not large. A few state groups have experience in R&D activities. Foreign enterprises investing in Vietnam are mainly in the fields of assembly, trade, and real estate. Therefore, there is little motivation to invent. Moreover, creating inventions with large-scale applications requires a huge investment in funding and personnel. Branches of multinational companies in Vietnam are only involved in manufacturing, distribution, and related activities, rather than developing new technical innovations. To create a premise for the role of enterprises in the NIS, Vietnam has set up a legal corridor for S&T enterprises to develop.

By 2007, the documents on the development of S&T enterprises had been separated from S&T activities and innovation with the introduction of Decree 80/2007/ND-CP dated 19 May 2007. They were dictated by government policies on S&T enterprises, followed by the program to support the development of S&T enterprises and S&T public institutions, and to implement the mechanism of autonomy and self-responsibility. The related laws are Decision No. 592/QD-Ttg dated 22 May 2012 of Prime Minister; adjustments in the Law on S&T (2013, Articles 57, 58), in the Law on Corporate Income Tax No. 32/2013/QH13 dated 03 June 2008; and Governmental Decree No. 91/2014/ND-CP dated 01 October 2014 (Mai Ha et al. 2015). These form the premise for businesses to enhance innovation activities.

From 2013 to 2016, the number of enterprises granted certificates for S&T enterprises increased from 87 to 250. They focused on areas such as biotechnology (47%), automation technology (17%), and new materials technology (14%). In addition, there were about 2100 enterprises that met the conditions of S&T enterprises. S&T enterprises have developed strongly and have started to participate in the technology supply system for the market, especially in high-tech sectors such as electronics and IT (64%), followed by mechanics and automation (19%), new materials (11%) and biotechnology (6%).[11] S&T enterprises are mainly concentrated in the big cities—Ho Chi Minh City and Hanoi.

5. *Investment funds for innovation activities in Vietnam*

S&T Fund: This fund is used to strengthen the S&T potential for businesses and industries, contribute to improving efficiency and competitiveness, implement the S&T tasks of enterprises, equip enterprises with material and technical facilities for business S&T activities, and purchase machinery and equipment along with technology transfer to replace part or all of the technology used with other more advanced technologies. The Model of Science and Technology Development Fund was mentioned in the Vietnam Law on Science and Technology in 2000. In addition to the national funds established by the government, ministerial-level agencies, governmental (ministerial) agencies, and provincial-level People's Committees set up S&T development funds to serve local activities and fields (Vietnam Ministry of Science and Technology, 2015).[12]

Vietnam has two funds with the most innovative S&T activities. The National Foundation for Science and Technology Development (NAFOSTED) and the National Technology Innovation Fund (NATIF). The year 2015 was a turning point in terms of changing the basic research funding method, the funding of potential/unplanned unexpected tasks, the program to support NAFOSTED's national S&T capacity building, proposed regulations on selection of S&T tasks, and the criteria for S&T experts under NATIF's database of experts.

According to NAFOSTED statistics, from 2009 to 2015, there were 3079 applications for registration and 1642 valid registration documents funded. The results of the basic research projects funded by the foundation showed that the research quality was maintained and accessible to international standards through the number of articles published in reputable and specialized scientific journals.

In 2014, under the Ministry of Home Affairs' Decision No. 1286/QD-BNV, issued on 16 December 2014, the Startup Vietnam Foundation was established. This is a venture capital fund model operating with capital mobilized from domestic and foreign investment sources to encourage the participation of Vietnamese organizations, citizens, and scientists with plans to start a business or invest in a business development based on the application of S&T.

Additionally, the local S&T Development Fund is established in provinces. The method of financial support by local S&T development funds is based on the following principles: (1) Partial funding for the implementation of research projects to create new technologies in the sectors and fields prioritized by the provinces, implemented by enterprises or in collaboration with S&T organizations (30% or less in the total funding for project implementation) and; supported to build quality management systems in accordance with international standards in enterprises (20% or less in the total funding for implementing projects on building quality management systems of enterprises); (2) Loans with interest rates lower than those of commercial banks at the time of lending or without bearing interest to implement projects that improve technology and apply S&T research results (trial production of new products and testing of new technology processes); and technology-transfer projects to innovate technology and products, convert production and business structure of enterprises, and serve the socioeconomic development objectives of provinces or cities (Nguyen Van Anh et al. 2012).[13]

3.1.1.2 Main Resources of the NIS

(a) *Human resources for research and deployment*

In the socialist-oriented market economy, S&T human resources are identified as an important development resource in the NIS. They determine the competitiveness and international integration ability of S&T. According to the Ministry of Science and Technology, in 2015, Vietnam had 167,746 participants in R&D activities, with some characteristics as described below (Charts 3.1, 3.2, 3.3 and 3.4):

Human resources for R&D were *mainly distributed in the state sector* (84.13%), while the non-state sector accounted for 13.8% and the FDI sector accounted for 1.07%. In each economic component (state, non-state, and foreign-invested), R&D personnel also accounted for the highest proportion.

Human resources for R&D were *mainly concentrated in universities, institutes, and colleges* (46.4%), followed by S&T research organizations (23.02%), mainly researchers (77.1% in S&T research organizations and 84.3% at universities, institutes and colleges).

The number of research human resources with university degrees accounted for more than 50% in the structure of research human resources. In the period 2011 to 2015, the proportion of researchers with postgraduate qualifications (doctoral and master's degrees) among researchers tended to increase, but slowly rose from 43.8% (2011) to 51.5% (2015). The proportion of female human resources accounted for about 45% of the total researcher population and tended to increase by 1.33 times in the period 2011 to 2015.

Research human resources with doctoral and master's degrees mainly worked in research institutions, universities, and colleges. Research human resources in enterprises accounted for about 14.8%.

Research human resources focused on S&T (34.9%) and social sciences (27.1%). In the fields of natural science, S&T, agricultural science, social sciences, and humanities, researchers were mainly concentrated in universities. In medicine and pharmacy, researchers focused more on the administrative agencies and non-business units. Research human resources in enterprises were mainly based in the S&T and social science sectors.

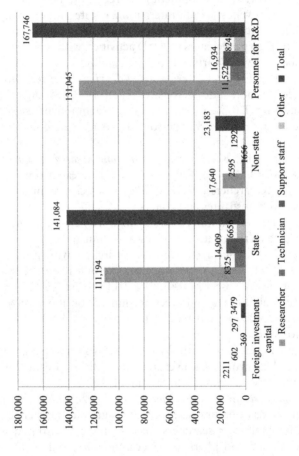

Chart 3.1 Research and deployment human resources by economic sectors and working functions. (Source: Vietnam Ministry of Science and Technology, 2016, p. 52)

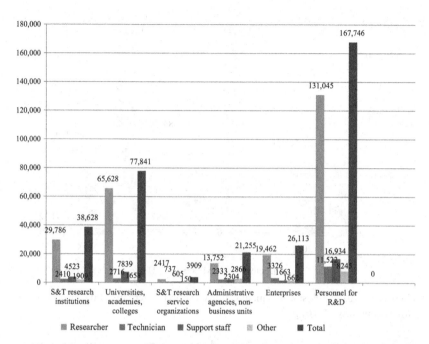

Chart 3.2 Research and deployment human resources by operational sectors and working functions. (Source: Vietnam Ministry of Science and Technology, 2016, p. 52)

Nowadays, there are a number of challenges facing S&T human resources for Vietnam's STI activities. The mobility of S&T human resources leads to gray matter flows. There is also an uneven distribution of S&T personnel (mostly concentrated in big S&T centers) as well as a lack of technical human resources and international integration capability.

(b) *Financial resources*

In Vietnam, financial resources for S&T comprise the following: state budget, investment from enterprises, and foreign capital. The structure of spending from the state budget for S&T has two main purposes: spending on investment and development and scientific career expenses. In the past, the state, in a centrally planned economy, managed all the resources. Since

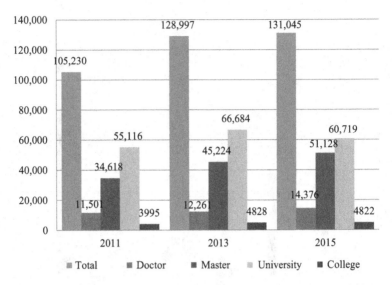

Chart 3.3 Research staff by qualifications. (Source: Vietnam Ministry of Science and Technology, 2016, p. 54)

Vietnam's move to a market-oriented mechanism, there have been many adjustments toward self-responsibility and the implementation of business accounting. The state supports and manages with policies and legal tools. The source of funding for S&T has been diversified. Particularly, the role of finance for S&T in enterprises is increasingly important.

- *Source of R&D funds for S&T from the state budget*

Sources of R&D funds for S&T from the state budget are invested in S&T organizations under the management of ministries, ministerial-level organizations, governmental organizations, national key laboratories, specialized laboratories, high-tech zones, and R&D funding for S&T in the provinces. In 2006 to 2016, most investments for S&T activities came from the state budget, accounting for 1.4% to 1.8% of the total annual state budget expenditure (excluding expenditures for S&T in national defense and security). However, this rate has tended to decrease in recent years. Investment from the state budget for S&T in 2016 reached VND17,730 billion, or 1.4% of the total state budget expenditure. The

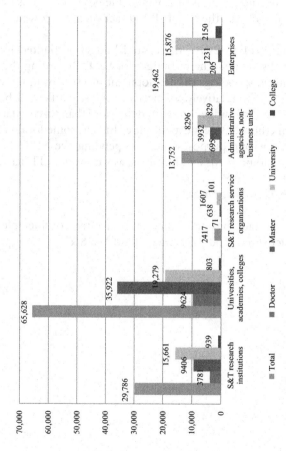

Chart 3.4 Research staff 2015 by qualifications and area of activity (per capita). (Source: Vietnam Ministry of Science and Technology, 2016, p. 55)

proportion of investment in S&T/GDP from Vietnam's state budget in the last five years was only approximately 0.4% (Fig. 3.1).

The total national expenditure for R&D activities tended to increase in the period 2011 to 2015 insignificantly (0.25%). Spending on R&D focused on science, engineering, and technology accounted for 71.84% of the total R&D spending. This is understandable because the business sector made up 63.6% of the total R&D activities, followed by the social sciences.

In the period 2011 to 2016, the central budget accounted for two-thirds of the total state investment budget for S&T. Regarding the structure of expenditure, spendings on development investment in the central and local budgets were relatively balanced. In the scientific non-business expenditure, the central budget accounted for more than three-quarters of the total scientific non-business expenditure. In fact, some localities used this fund for the wrong purposes such as spending the S&T funds on investments in building roads and hospitals, as well as pure IT activities.

- *Enterprise investment for S&T*

Along with the state, businesses have also invested considerable funds in their S&T activities. The investment fund for S&T in enterprises is

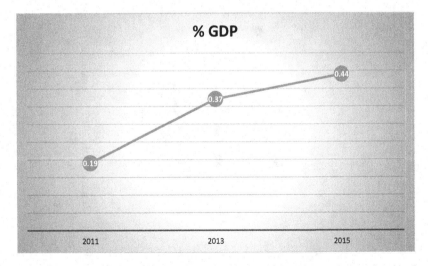

Fig. 3.1 Gross domestic spending on R&D. (Source: Vietnam Ministry of Science and Technology, 2016, p. 61)

divided into two groups: R&D activities and technology renovation activities.

- *Enterprises tend to increase funding for R&D activities:* In the period 2012 to 2015, investment in R&D activities in this sector tended to increase. According to the data from the R&D survey in 2012 by the General Statistics Office, the expenditure on R&D reached VND1500 billion, accounting for 27.58% of the national funding for R&D. In 2013, the investment of enterprises for R&D was VND5597.3 billion (accounting for 41.8%) (Vietnam Ministry of Science and Technology, 2014).[14] In 2015, this value was VND11,766.2 billion and accounted for 63.61% (Vietnam Ministry of Science and Technology, 2016).[15]

(c) *Information resources*

In 2000, official STI activities were mentioned in the Law on Science and Technology passed by the National Assembly on 9 June 2000.[16] For the first time, STI activities in Vietnam were determined by law, which indicated:

> The Government invests in the construction of a modern national scientific and technological information system, ensuring adequate and timely information on important achievements in domestic and international scientific and technological fields; promulgating regulations on management of scientific and technological information. (Article 45)

According to the 2016 Science and Technology White Paper, by the end of 2015, there had been a total of 334 scientific journals with International Standard Serial Number (ISSN) included in the review list by the State Council for Professorship Title. Among them, *Advances in Natural Sciences: Nanoscience and Nanotechnology* (ANSN), published by the Vietnam Academy of Science and Technology in collaboration with IOP Publishing in the United Kingdom, was indexed in ISI/SCIE (Institute for Scientific Information/Science Citation Index Expanded) in January 2016, after being included in the Scopus database since 2014.

In addition, Vietnam has two journals in mathematics classified in the Scopus database. They are *Acta Mathematica Vietnamica*, published by the Institute of Mathematics since 2011, and *Vietnam Journal of Mathematics*, published by the Vietnam Mathematical Society and the Vietnam Academy of Science and Technology since 2014. Of the 334 scientific journals mentioned above, only 26 (0.078%) were published in English wholly or in part. The whole S&T network currently has over 5 million copies of S&T books, and access to more than 20,000 online S&T journals with over 40 million full-text records. Among them, S&T journals are primarily provided online through VISTA and VinaREN Networks of the National Department of Science and Technology Information. In order to meet the demand of online exploitation and sharing of S&T information, Vietnam Journals Online-VJOL is maintained and expanded, allowing readers to access scientific knowledge published in Vietnam and improve the world's knowledge of Vietnamese scholarship.[17]

Electronic information resources continue to see strong subscription, from domestic databases to the world's leading databases such as Science Direct, Proquest Central, Web of Science, IEEE, APS, Primo Central Index, IOP Science, Springer eJournals, and so on. There are currently more than 1000 databases built by S&T information agencies, in which nearly 10% of databases have more than 10,000 bibliographic records each. In particular, there are a number of large databases with hundreds of thousands of records such as the Database of Vietnamese Science and Technology Documents with over 200,000 records allowing access to full-text documents and databases of S&T books.

3.1.1.3 Some Results of STI Vietnam System

(a) *Number of S&T publications*

- *Domestic publications*

    - *Growing number of domestic articles*: According to statistics from the Vietnam Database of Science and Technology, which collects S&T publications from 236 S&T journals (accounting for 70% of the total number of domestic S&T journals), Vietnam has about 220,000 scientific papers published domestically. From 2011 to 2016, the number of domestic articles increased 1.74 times.[18]
    - *Fields of research*: In 2016, articles focusing on social sciences accounted for the highest proportion (57.53%), science, engineering, and technology (15%), humanities sciences (13.5%), medical science (5.19%), and natural science (4.38%). Although Vietnam is known as a country with traditional agriculture and

a lot of agricultural products, the number of domestic research and publication in the agricultural sector is the lowest, at 4.40%.

- *International publications*

  - Growth rate of articles: Between 2001 and 2005, Vietnam had 2497 international scientific publications and was ranked 73rd in the world in terms of the number of scientific publications published in the Web of Science database. By 2006 to 2010, the number of international publications had increased to 5228 articles and Vietnam was ranked 63rd in the world. In the period 2011 to 2016, the total number of S&T publications in Vietnam in the Web of Science database was 16,104 articles, ranking 4th in the total number of international publications by ASEAN countries (after Singapore, Malaysia, and Thailand). However, Vietnam's number of international publications is only one-third that of Thailand, which had 49,750 articles.[19]
  - *Fields of publications*: In the period 2011 to 2018, Vietnam's international publications in the Web of Science database focused mainly on the fields of physics (2034 articles, or 12.6%), technology (1856 articles, or 11.5 %), and mathematics (1770 articles, or 11%). In 2018, Vietnam ranked 46th in the league of ISI publications in physics, 40th in engineering, and 32nd in mathematics.[20] Publications in fields such as agriculture, biochemical molecular biology, applied microbiology, biotechnology, food technology, microbiology, business and economics, and immunology accounted for less than 3% of the total records of Vietnam's international publications in the same year.
  - *Citations of Vietnamese articles*: Although there were more articles published in the period 2011 to 2018, the number of citations of Vietnamese articles tended to decrease. Possible reasons may include: (1) the novelty and reliability of the research results are not high; (2) research issues are not really of the interest in the international community.

(b) *Registration of inventions and utility solutions*

According to the Department of Intellectual Property, the registration of inventions and utility solutions accounted for about 11% of industrial property applications filed directly in Vietnam in 2017 (Chart 3.5).

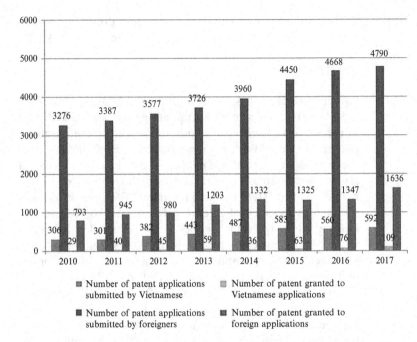

Chart 3.5 Data on the filing of patent applications and patent numbers of Vietnamese and foreigners filed in Vietnam, 2010–2017 (Source: National Office of Intellectual Property of Vietnam, 2017, p. 70)

- *Patent registration*

    – The number of patent applications increased by 1.5 times from 3582 in 2010 to 5382 in 2017. Accordingly, the number of patents also increased by 2.12 times from 822 patents in 2010 to 1745 patents in 2017. However, the number of patent applications and patents granted were mainly from foreigners. The number of applications and patents by Vietnamese accounted for a very low percentage (fluctuating within the 3% to 10% range) over the years. One of the reasons is that the technical solutions of these inventions are not new. The level of creativity and ability to apply for patents and inventions in industries may not be of high practicality or quality. The quality of Vietnamese registration applications fails to be sufficiently informative or demonstrate the superiority of the invention.

- *Utility solutions*

In the period 2010 to 2017, the number of applications for utility solutions increased from 299 to 434, by 1.45 times. The number of granted patents of utility solutions also increased from 58 in 2010 to 146 in 2017, by 2.5 times. The applications and patents were mainly from Vietnamese people (accounting for 50%–70% of the total number of applications and patents) (Chart 3.6).

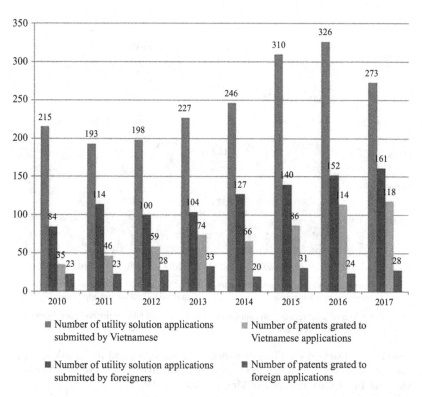

Chart 3.6 Data on the application for registration of utility solutions and number of patents of useful solutions of Vietnamese and foreigners submitted in Vietnam, 2010–2017. (Source: National Office of Intellectual Property of Vietnam, 2017, p. 72)

In this period, the number of trademark registration applications also increased from 27,923 in 2010 to 43,970 in 2017, by 1.57 times. The number of applications accepted increased from 16,520 in 2010 to 19,401 in 2017, by 1.17 times. Foreign applicants made up the majority, but the applications accepted were mostly from Vietnam.[21] Regarding industrial designs, the numbers of applications filed and granted patents for industrial designs tended to increase and were mainly granted to Vietnamese (Charts 3.7 and 3.8).

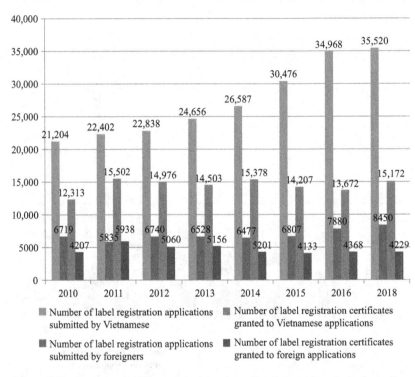

Chart 3.7 Data on the filing of trademark applications and trademark applications of Vietnamese and foreigners filed in Vietnam, 2010–2017. (Source: National Office of Intellectual Property of Vietnam, 2017, p. 94)

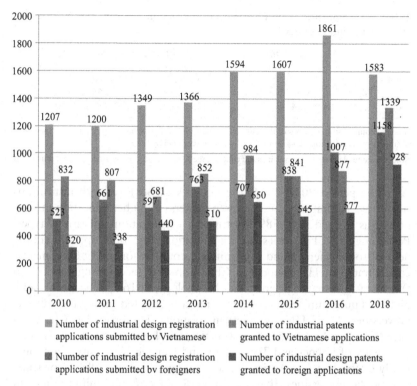

Chart 3.8 Data on the filing of industrial design registration applications and the number of applications for industrial design registration of Vietnamese and foreigners submitted in Vietnam, 2010–2017. (Source: National Office of Intellectual Property of Vietnam, 2017, p. 92)

3.1.2 Current Status of STI Activities in Vietnamese Universities

In Vietnam, 90% of S&T human resources throughout the country are provided by domestic universities, and 10% are trained abroad. S&T human resources not only directly participate in S&T activities, but also play management and leadership roles in state management agencies for S&T, S&T organizations, and enterprises. Furthermore, national universities, regional universities, and private universities are multidisciplinary. Public universities account for over 70% of the total number of universities in Vietnam (Table 3.1).

Table 3.1   Statistics of universities and colleges in Vietnam for 2014–2015, 2015–2016, and 2016–2017

	2014–2015	*2015–2016*	*2016–2017*
Public	156	163	170
Non-public	58	60	65
Total	214	223	235

Source: Summary from statistics of the Ministry of Education and Training, Website: https://moet.gov.vn/thong-ke/Pages/thong-ko-giao-duc-dai-hoc.aspx?ItemID=5137

Note: Of the total number of universities and colleges, the schools of security and defense are not included

In the period 2011–2016, the network of S&T research institutions (institutes, laboratories, and research centers) in higher-education institutions was formed with 589 institutional organizations of various types. Pertaining to S&T activities, higher-education institutions operating in the field of social sciences and humanities accounted for nearly 50%, mainly concentrated in Hanoi and Ho Chi Minh City (accounting for 27.4%) (Vietnam Ministry of Science and Technology, 2016).[22] The R&D capabilities of public universities were mainly concentrated among the national universities (VNU-HN and Vietnam National University Ho Chi Minh City (VNUHCM)), regional universities, and leading universities (Hanoi University of Science and Technology and Danang University). For some newly established universities that were upgraded from colleges, or some non-public universities, the capacity of R&D was still limited (Vietnam Ministry of Education and Training, 2017).[23]

3.1.2.1   S&T Human Resources in Vietnamese Universities
In 2016, the statistics by the Ministry of Science and Technology showed that Vietnam's S&T human resources were concentrated in the university area, with 77,841 people, accounting for 46.4% of the total R&D personnel (167,746 people).[24] Among them, the R&D human resources of the university mainly focused on researchers (65,628 people, or 84.3%), technical personnel (2716 people, or 3.48%), assisting staff (7839 people, or 6.1%), and others (1658 people or 2.12%). In particular, the number of R&D personnel with master's degrees was the highest at 35,922 people (54.7%), followed by those with university degrees (19,279 people, or 29.37%), while 9624 people (14.66%) held doctoral degrees. S&T human resources at universities were concentrated in the social sciences (21,396 people, or 32.6%), as well as science, engineering, and technology (19,280 people, or 29.3%). The least number of employees worked in medical and

pharmaceutical science (5513 people, or 8.4%) and agricultural science (4410 people, or 6.71%).

The two national universities had the highest number of teaching and research personnel among the existing universities. Vietnam National University Ho Chi Minh City (VNUHCM) had the largest number of teaching and research personnel with doctorates and scientific doctorate degrees, as well as professor and associate professor titles. This is entirely consistent with the goals of the two national universities to become the leading research universities in the country. At most of the universities, most lecturers mainly teach and participate in research for teaching, or technology development associated with consulting activities. The Hanoi University of Science and Technology is one of the strongest institutions in S&T teaching and research. Regional universities such as Thai Nguyen University also have a relatively high number of scientific personnel, compared with the remaining three institutions that have the lowest numbers, but there is a marked change in quantity and quality.

The research team collected data between 2017 and 2018 at four universities: VNU-HN, Vietnam National University Ho Chi Minh City (VNUHCM), Hanoi University of Science and Technology, and Thai Nguyen University. These are representatives of the university sector and are considered to be highly promising in scientific research and innovation activities. In all four universities, only the Hanoi University of Science and Technology has a higher number of human resources with doctoral degrees than the number of master's degree holders, while in the other universities, the number of master's holders remains dominant (Chart 3.9).

3.1.2.2 Investment in STI Activities

According to the 2015 statistics by the Ministry of Science and Technology, the source of financial investment received by universities in Vietnam was mostly from the state budget (VND1015.9 billion), accounting for 95.5%, and the budget from non-state and foreign investment capital was VND47.3 billion, accounting for only 4.4% of the total budget for R&D activities (VND1063.2 billion). The financial investment for S&T was VND468.6 billion (44.07%), and the lowest was in humanities with VND50.7 billion (4.7%).

In terms of the expenditure for various types of R&D activities, the university sector ranked second after hospitals and research centers (Table 3.2).

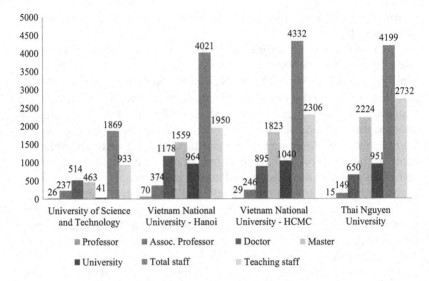

Chart 3.9 Number of officials, lecturers, and staff qualifications of four universities, 2017–2018. (Source: Thai Nguyen University, 2018 and Institute of Policy and Management, 2019)

According to 2014 statistics published by Vietnam National University Ho Chi Minh City (VNUHCM) (Table 3.3), its total funding for S&T was VND362.798 million, and VND149.902 million, or 41%, of this amount was from the central budget. The fund mobilized from domestic and foreign cooperation was VND58.796 million, or 16%. The technology-transfer fund was VND154.100 million, or 43%.

At VNU-HN, the total number of cooperative funding for the implementation of domestic collaborative S&T projects reached more than VND38 billion (nearly 25% of the fund for S&T activities). In addition, VNU-HN organized and approved 20 state-level S&T tasks under the Northwest Program in 2013 and 2014, which saw scientists from the university successfully bid for six tasks with a total cost of VND51.8 billion.

Funding for S&T activities at the Thai Nguyen University included the state budget, tuition fees, and contracts with scientific research projects with local enterprises. The main source of funding still came from the state budget, but in 2014 this source of funding was greatly reduced (only a third of the previous years').

Table 3.2 Expenditure for R&D by type of activity (excluding business sector) (billion VND)

Type of activity	Area of implementation				
	Institute, research center	University	Administrative, non-business	Non-business unit	Non-state
Basic research	807	309	43	27	8
Application research	1.167	376	253	244	28
Experimental implementation	235	49	69	82	24
Testing production	104	24	28	30	8
Total	2.313	758	393	383	68

Source: Ministry of Science & Technology, Science and Technology of Vietnam (2013, p. 198)

Table 3.3 Budget of Vietnam National University Ho Chi Minh City (VNUHCM) from investment sources for S&T research in 2014 (*Unit: million VND*)

Budget resources	Budget granted in 2014 (million VND)
1. From central budget	149.902
2. From cooperation	58.796
Local cooperation	*44.071*
Enterprise and international cooperation	*14.725*
3. Technology transfer	154.100
Total (1 + 2 + 3)	362.798

Source: Report on S&T research activities, Vietnam National University Ho Chi Minh City (2014, p. 135)

The Hanoi University of Science and Technology obtained funding for scientific activities mainly from the state budget (Table 3.4). Besides, this university had funding from production cooperation activities with businesses (as of 2008, the university's revenue generated from cooperative activities with enterprises was VND2904.45 million).

In summary, the main source of funding for S&T activities in these universities is from the state budget. This allocation of funding differs between the national universities and regional universities (such as Thai Nguyen University). It must be said that this funding for S&T activities is too small to encourage the implementation of S&T innovation. Activities only stop at the implementation of research topics at the levels ordered.

Table 3.4 Funds for implementation of S&T activities at Hanoi University of Science and Technology, 2011–2015 (*Unit: million VND*)

Topics of level	Total funding	Total funding support from the state budget
State	311,980	311,980
Protocol	4,032,316	26,316
Ministry	104,380	104,380

Source: Science and Technology Report of Hanoi University of Science and Technology, 2006–2015 (p. 136)

Active research based on assistance funding or sponsorship is very limited. Therefore, it is inevitable that the effectiveness or results of scientific research activities is not high. These institutions simply carry out the scientific tasks that they have been ordered to do. Their scientific research spirit is not highly appreciated. However, these universities are currently making efforts to be self-reliant and seek funding sources from cooperation and technology transfer from enterprises and investment funds. To improve their S&T activities, they are calling investment participation from networks of domestic and foreign S&T enterprises as well as organizations to diversify investment capital sources.

3.1.2.3 Organization of STI System in Universities

The STI system in these universities includes member universities or affiliated faculties, member scientific research institutes, research, service and training centers, and units for research and training activities.

Specifically, Vietnam National University Ho Chi Minh City (VNUHCM) has six member universities, one research institute, one affiliated department, and some research and service centers (English Testing Center, Training Quality Testing & Assessment Center, National Defense—Student Security Education Center, French University Center, and Integrated Circuit Design Research and Education Center). VNU-HN has seven member universities, five affiliated faculties, five member scientific research institutes, four research, service and training centers, and ten units for research and training activities (Dao Thanh Truong, 2016b). Hanoi University of Science and Technology has 4 training and research centers, 21 member science institutes, 8 affiliated companies, and some units for training and research activities. Thai Nguyen University has three research institutes, one university hospital, eight centers for S&T research, and one limited liability company.

Among the four research universities, there are two universities with affiliated companies: Hanoi University of Science and Technology and Thai Nguyen University. The former is one of the first universities in Vietnam to establish a business incubator model and holds the most affiliated enterprises among the four. It must be said that the establishment of institutions with technology incubation functions in universities has been a big step in integrating research, training, and production functions in the STI system. This model has brought about efficiency to improve the capacity of inputs and the efficiency and quality of outputs in S&T activities. This would create a spreading effect on other universities in the future.

According to a survey of 142 out of 271 universities that was conducted by the Ministry of Education and Training in 2017, the university system formed 945 research groups, and each university had 7 research groups on average. The number of S&T research groups is uneven. Hanoi University of Science and Technology, Hanoi University of Science, VNU-HN, University of Agriculture and Forestry, Thai Nguyen University, Hanoi National University of Education, Hanoi University of Science and Technology, and Danang University had the largest number of strong research groups.[25]

3.1.2.4 Some Results of the STI Systems of Vietnamese Universities

The state budget for investment in S&T activities in the field of education for the Ministry of Education and Training increased an average of 8% per year during the period 2006 to 2012. The number of universities has also grown. S&T organizations, including universities, have received a lot of state investment since they are an important component in research activities and create a majority of intellectual property to promote innovation in Vietnam. Currently, the government also supports many universities in Vietnam to build and pursue research university models. Training, scientific research, and technology transfer are considered the three pillars of research universities. In fact, the number of products and research results from universities has increased in recent years. According to the 2011 preliminary statistics from the Ministry of Science and Technology, about 2000 S&T tasks were implemented by S&T organizations such as universities, colleges, and institutes/research centers.[26] Each year, universities train about 15,000 master's students and 1000 doctoral candidates. The university sector contributes about 16,000 to 18,000 research results annually.

In September 2014, according to the SCImago Institutions Rankings, out of 2744 universities and research institutes and institutions worldwide, Vietnam had four on the list. They were the Vietnam Academy of Science and Technology, Vietnam National University Ho Chi Minh City (VNUHCM), VNU-HN, and the Hanoi University of Science and Technology.

In the ranking of the top universities in Asia in 2019 by Quacquarelli Symonds, seven universities in Vietnam, including VNU-HN, Vietnam National University Ho Chi Minh City (VNUHCM), Hanoi University of Science and Technology, Ton Duc Thang University, Can Tho University, Hue University, and University of Da Nang, were ranked. The ranking by Quacquarelli Symonds—QS-Asia[27] was based on the following criteria: (i) academic reputation and training (30%); (ii) reputation of the recruited person (recruited graduates) (10%); (iii) proportion of lecturers/students (20%); (iv) coefficients of citation/scientific articles (15%); (v) number of scientific articles/lecturers (15%); (vi) proportion of international lecturers (2.5%); (vii) proportion of international students (2.5%); (viii) rate of exchange students in the country (2.5%); and (ix) the percentage of foreign exchange students (2.5%).

Within a dynamic economic environment, Vietnam National University Ho Chi Minh City (VNUHCM) made efforts to cooperate with many leading universities and research institutes to implement research and training programs on human resources in scientific fields that are both modern and applicable in society. Typically, there have been projects of cooperation with MINATEC (Micro and Nanotechnology Innovation Centre), INPG (France), and the University of California at Los Angeles. In addition, this unit cooperated very closely with five domestic groups and corporations in order to strengthen the output and input capacity of the quality of human resources as well as the effectiveness of S&T activities. In 2014, Vietnam National University Ho Chi Minh City (VNUHCM) published 3038 conference papers/reports on all fields. The number of articles published in domestic and foreign journals was 1147 articles, 48.5% (566 articles) of which were published in international journals, and 30% (341 articles) in ISI journals. Among them, 46% (128 articles) had an impact factor greater than 2; 82% (281 articles) had main co-authors based in Vietnam National University Ho Chi Minh City (VNUHCM) and 47% (161 articles) had sole authors based in Vietnam National University Ho Chi Minh City (VNUHCM). Also, according to the S&T activity report, the total number of patent applications reached 85 in 2014. In particular, patent protection was granted for 78 applications and more than 100 S&T products.

Despite being a regional university, the results of S&T activities that Thai Nguyen University has achieved are very remarkable. In 2014, Thai Nguyen University chaired and implemented 10 state-level S&T tasks (including 2 independent projects, 3 research projects under the protocol, 2 biotechnology application projects, and 5 basic research projects funded by NAFOSTED) and 51 ministerial-level scientific tasks, including 47 S&T research projects, 01 S&T task to support the management and direction of the industry, 01 task of genetic fund, 03 pilot production projects, and 01 research capacity enhancement project. Regarding cooperation activities in research, Thai Nguyen University conducted two research projects under the protocol, one biotechnology application program, some research projects of bilateral cooperation, especially programs on economic development and social culture in the northern midland and mountainous provinces with a total cost of nearly VND71 billion. Based on the unit's report in 2014, Thai Nguyen University published 904 articles in domestic scientific journals and 177 articles in international journals. The University of Science (Thai Nguyen University) was the unit with the most articles published internationally, with 43 articles, 31 of which were in the ISI list.

Through research and practical activities, the Hanoi University of Science and Technology is one of the pioneering institutions of S&T activities in the university sector within the STI system. In the period 2011 to 2015, the Hanoi University of Science and Technology conducted 116 state-level projects with 22 topics and cooperated with more than 10 enterprises and corporations in the S&T field.

Technological transfer in universities has recently achieved remarkable results, promoting socioeconomic development in the context of regional and global economic integration. Some universities have achieved high turnover from commercialization of their inventions. Like the Hanoi University of Science and Technology, many technology-transfer contracts have been signed with many big organizations in Vietnam and globally, such as SUN MicroSystems Group, Rang Dong Water Bottle, and Bulb JSC. The university focuses on research, technology development, and human resource training for scientific research. Within the overall university block, only a few universities have achieved success in technology transfer activities in particular and commercialized research results in general. They mostly focused on S&T sectors. Comparatively, however, this number is not sufficient to meet public needs. It is still at potential levels in both quality and quantity compared to other components in the chain of S&T system links. In developed countries, meanwhile, universities and research institutes play an important role as a bridge to provide input resources (S&T human resources, S&T information) and create an environment in the nursery stage for the next S&T activities. This is an extremely important step in the STI activity series. The close connection with other components such as enterprises or research institutes not only helps universities improve the quality of S&T human resource training, but also promotes the applicability of scientific research results and technology development.

In many universities, however, research activities are still sparse. The topics have not yet brought about high practical value and are mainly topics and projects under orders from the state with few orders from manufacturing enterprises. Furthermore, the management of intellectual property in universities has not been given adequate attention. This results in the university's technology transfer activities not attracting participation from human resources, and there is no organization with sufficient professional capacity to manage and put the transfer into production and business. In the context of teachers and scientists connecting with scientific research activities under limited resources, the cooperation between universities and enterprises becomes an important solution. Not only are universities providing human resources, transferring the results of scientific research into production, and creating an ideal 'practicing' environment for students, but also benefiting from profits and increasing competitiveness for businesses by putting into production and applying scientific research results. Even now, most universities have specialized departments to manage scientific research and technology transfer activities or the establishment of technology incubators. However, these tasks have not been implemented effectively.

From the time of the former Soviet Union, Vietnam and the other socialist countries were in a similar situation: universities were primarily assigned to do the task of training. Scientific research was left almost unattended and considered as a separate function under the responsibility of academia. As a result, universities only participated in research within the framework of programs and topics of the state at all levels, and those involved in scientific research were part-time, not professional. This inhibited the development and connection of scientific research and training functions in universities.

In Vietnam, the issue of the 'research university' has emerged. Universities and society increasingly need to connect research and training functions in order to strengthen links with other components such as research institutes and businesses. The value chain in Vietnam's STI system is being formed and requires all components to strive to develop endogenous capacity and develop exogenous relationships to develop the system and adapt to the trend of international S&T integration. However, general regulations for research universities today are not yet available. These are only applied at the Hanoi University of Science and Technology and VNU-HN.

3.1.3   Current Status of STI Activities of Research Institutes

3.1.3.1 Overview
A sociological survey was carried out in 2014 by a research group in the state-level scientific project, KX06.06/11-15. The team issued more than 200 sheets, and 120 sheets were collected. Among these, there were 60 sheets for research institutes and 60 sheets for individuals working in research institutes to find out more about the STI system in research institutes, and the difference between viewpoints and institutional opinions pertaining to the organization of the STI system.

- *S&T human resources of institutes*

The biggest challenge for Vietnamese S&T today is that the team of qualified scientists is less able to meet the needs of economic and social development. But the revised S&T law does not have specific provisions to help build a comprehensive and stable development team.

According to the survey data on the structure of R&D human resources in the research institutes, 91% were research human resources, only 4% were technical personnel, and 4% were support staff (Chart 3.10). Although this was a human resource structure in the research institutes, the percentage of technical personnel and support staff was quite low (Table 3.5).

Data shows that there were no foreign workers, so the table does not include this. In terms of gender in the human resource structure, among the 60 research institutes, there were about 40.5 female research workers in one institute, accounting for 87.7%. An average of 2.3 female technical personnel accounted for 5%. There was an average of three female support staff working in research institutes, accounting for 6.5%, and an average of 1.4 people in other capacities, accounting for 0.8%.

According to Table 3.6, the average number of doctoral-level staff in each institution was 28.2 men, and that of master's degree holder was 19.6 men. In terms of titles, each institute had an average of 2.3 male

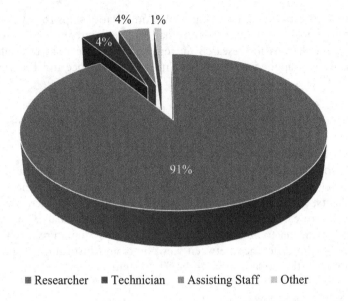

4%   1%

4%

91%

■ Researcher   ■ Technician   ■ Assisting Staff   ■ Other

Chart 3.10   Structure of R&D human resources in research institutes. (Source: Survey data from the project KX06.06/11-15)

professors and 8.4 male associate professors. Meanwhile, the number of female doctors, female master's degree holders or female professors is less than half of that of males with equivalent qualifications and titles. Even at the professor level, there were no female professors taking part in S&T activities. The number of highly qualified female manpower in research institutes is lower than that of men because women who pursue and do scientific research especially at the professor level are quite few, and there are many constraints that limit them to a certain professional level. In 2013, Vietnam had 10,453 professors and associate professors. The percentage of women with master's degrees accounted for 30.53%, those with doctoral degrees accounted for 17.1%, those with assistant professorship accounted for 11.7%, and those with professorship accounted for 5.1%.

In a nutshell, S&T human resources at institutes need to be further developed. Because of the current situation, young people do not want to pursue a scientific career because they do not see the future of development and a stable life. In order to attract the young elite to choose a scientific path, the state must have wage policies to ensure that scientific

Table 3.5 Average number of human resources by specialties in S&T activities of research institutes

Manpower	Vietnamese	
	Average value	
	Male	Female
	Mean	Mean
1. Research human resources (with college degree or higher), spending at least 10% of their time for R&D	69.8 people	40.5 people
2. Technical personnel (including technicians, laboratory staff with intermediate and equivalent qualifications)	2.9 people	2.3 people
3. Support staff (administrative and office staff)	1.4 people	3 people
4. Others	0.9 people	1.4 people

Source: Survey data from the project KX06.06/11-15 (Dao Thanh Truong, 2016a)

Table 3.6 Average number of human resources by educational background in S&T activities of research institutes

Nhân lực	Vietnamese	
	Average value quantity	
	Male	Female
	Mean	Mean
1. Divided by professional level		
Doctor	28.2 people	12 people
Master's degree	19.6 people	14.2 people
University degree	19.8 people	11.4 people
College degree	1.3 people	1.4 people
2. Divided by academic title		
3. Professor	2.3 people	0 person
4. Associate Professor	8.4 people	1.5 people

Source: Survey data from the project KX06.06/11-15 (Dao Thanh Truong, 2016a)

human resources have stable income and living conditions. It is necessary to adjust the regulations on current starting salaries of researchers as well as implement supporting policies for scientific human resources. Without a breakthrough in S&T human resource policy, there will certainly not be enough S&T human resources in the future.

R&D spending has always been considered a leading R&D indicator. It is regarded as the main indicator for evaluating a country's R&D strength. Of the total number of research institutes participating in the survey, the expenditure on S&T activity in the research institutes accounts for only VND414.5 billion, and each research institute spends VND6.9 billion on S&T activities.

Among them, *capital from the state budget* generally accounts for a large proportion with 70% to 100% of the institute capital. More specifically, there are 10 research institutes that have 70% capital from the state budget (16.7% of the 60 institutes studied). The state budget accounts for 75% of the S&T spending at 10 research institutes (16.7%), and there are 10 research institutes (16.7%) with 100% funding for S&T activities from the state budget.

Research institutes that use their own funding for R&D activities include 10 research institutes (16.7%), and their funding accounts for 5% of the institute budget.

There are 10 research institutes (16.7%) spending 20% of the budget. There are another 10 research institutes (16.7%) that spend 100% on R&D activities.

Enterprise capital expenditure for R&D activities of institutes has a relatively low rate compared to the private capital of enterprises. There are 20 research institutes (33.3%) receiving funding from businesses for 10% of their R&D costs. There are 10 research institutes (16.7%) that receive 20% funding for their R&D expenses. This proves that the association between research institutes and enterprises is still not tight. This may be due to reasons such as the incompatibility of the institutes' research for enterprises, or enterprises' funding for R&D activities is limited.

The survey results on funds for S&T activities of institutes found that there was no funding from the research institutes themselves and from non-governmental organizations. Besides, 0.7% of research institute funding was collected from copyright sale. It can be seen that the research institutes were mainly funded by the state budget. Funding from businesses had quite a modest amount of capital compared to the private capital of the institute.

Among the research institutes participating in the survey (Table 3.7), 20 of them (33.3%) spend between 30% and 70% of their budget on R&D, and 40 research institutes (66.7%) spend more than 70% of their budget on R&D.

In terms of *research institutes spending on S&T services*, 91.7% of the institutes spend less than 70% on S&T services. Among them, 40 research

Table 3.7 Ratio of fund allocation for S&T activities of research institutes

		Quantity	%
Funding for research and deployment	Less than 30%	0	0
	30–70%	20	33.3
	Over 70%	40	66.7
	Total	60	100.0
Funding for S&T services	Less than 30%	15	25.0
	From 30–70%	40	66.7
	Over 70%	0	0
	No answer	5	8.3
	Total	60	100.0

Source: Survey data from the project KX06.06/11-15 (Dao Thanh Truong, 2016a)

institutes (66.7%) spend 30% to 70% of their funds on S&T services. Another 15 institutes (25%) spend less than 30% of their funds. Five institutes did not provide an answer. It can be seen that the proportion spent on S&T services is not reasonable for developing S&T integration with the world.

Besides, the research institutes also pay for other aspects such as human resource training and land lease.

Training cost for S&T human resources: The research institute is a highly qualified human resource training institution. However, 30, or half, of the research institutes only spend less than 30% for training. There are 20 institutes (33.3%) spending 30% to 70% on training activities, and 10 research institutes (16.7%) spend 70%.

Average cost of S&T manpower: There are 23 research institutes, or 38.3%, spending less than 30%, 14 research institutes (23.4%) spend between 30% and 70%. There are 23 research institutes (38.3%) spending over 70%. This is understandable because, according to 2011 data, a research worker is paid an average of USD5857 per year in Vietnam, while in Indonesia in 2009 the amount was USD19,530 (three times higher than in Vietnam), and Malaysia (2006), USD109,905 (20 times higher than in Vietnam).[28] Hence, the average cost of S&T human resources in Vietnam is quite low in ASEAN.

Funding for basic research: 50% of the research institutes have a cost ratio concentrating mainly at the 10% to 30% range; 25% of research institutes spend 31% to 70% on R&D, and none of them spend more than 70% on R&D activities. Compared to other types of research, the level of spending on basic research is only average, but still higher than the level for experimental implementation. This survey data is similar to the 2013 data by the Ministry of Science and Technology, which announced that research institutes' spending on basic research was VND807 billion, 1.5 times less than the spending on applied research (VND1167 billion), and 3.4 times more than experimental implementation.

In terms of the amount research institutes spent on applied research, most institutes spend 31% or more. There are 15 institutes spending over 70%, and 30 institutes spending 31% to 70% for applied research. It can be seen that applied research attracts a large amount of investment funding. According to data published in 2013 by the Ministry of Science and Technology, the exclusive funding for the institutes' applied research was VND1167 billion, 3 times higher than that of universities (VND376 billion) and nearly 42 times higher than that of non-state units (VND28 billion). Compared to basic research (VND807 billion), the expenditure on applied research at research institutes was 1.5 times higher, 5 times higher than experimental implementation (VND235 billion), and 11 times more than trial production (VND104 billion). Thus, funding for applied research is the highest compared to other types of activities.

Research institute expenditure for experimental implementation: 73.4% of the institutes spend less than 70% on experimental implementation. Twenty-four institutes (40%) spend less than 10% on experimental implementation. Ten research institutes (16.7%) spend between 10% and 30%, and another 10 research institutes (16.7%) spend between 31% and 70%. Experimental implementation constitutes a low expenditure in the budget of the institutes. In fact, the budget spent on experimental implementation by research institutes is the third largest in the four R&D activities (basic research, applied research, experimental implementation, and trial production). Funding for experimental implementation is only one-fifth that of applied research, which is equivalent to a quarter of basic research and two times of trial production.

Opportunities of research institutes receiving venture capital: In Vietnam, the problem of the lack of capital for S&T development to meet economic growth needs becomes a vicious cycle: less capital leads to low investment in S&T, which results in low productivity and low growth. In turn, the low growth rate inevitably leads to low levels of capital accumulation for the economy. Recently, Vietnam has revised state policies to facilitate the diversification of financial resources for investment activities, with 75% of research institutes receiving venture capital. However, 25% of research institutes did not receive this funding. Capital sources and venture capital funds in Vietnam are limited, although this is an essential activity for S&T development.

- *Organization of research institutes' STI system*

Besides the training of human resources, scientific research and technology transfer are among the vital tasks of a research institute. In assessing the institutes' activities of S&T organization, the author's research team used the results of 60 sheets for individuals operating in the institute out of 120 sheets for research institutes (Table 3.8).

The table shows that most researchers evaluating the state's incentives for STI activities at an average level account for more than 70%. In particular, the percentage of the respondents working for 10 years or more who evaluate the factor as average was 100%. The 31.6% of senior respondents working for less than 10 years evaluate that the state's incentives in this activity are still weak. The average level accounts for 68.4%.

Regarding the assessment of human resources for innovation activities, there is a difference between the group of respondents who have worked for 16 years or more and the other two groups. The group working for 16 years or more evaluates "human resources available in S&T at all times" only at the weak and average levels, with 71.4% (five people) evaluating as weak and 28.6% (two people) evaluating as average. On the other hand, 73.7% of the group working less than 10 years evaluates this factor at a

Table 3.8 Assessment of state incentives and human resources for R&D activities and innovation activities in Vietnam today by number of years of service

		Less than 10 years		10–15 years		16 years or more	
		Quantity	%	Quantity	%	Quantity	%
State incentives for	Poor	6	31.6	0	0	0	0
innovation available at	Average	13	68.4	34	100.0	7	100.0
all times	Good	0	0	0	0	0	0
Human resources	Poor	0	0	4	9.1	5	71.4
available in S&T at all	Average	5	26.3	10	29.4	2	28.6
times	Good	14	73.7	20	61.5	0	0

Source: Survey data from the project KX06.06/11-15 (Dao Thanh Truong, 2016a)

Table 3.9 Assessments on the management of institutes related to S&T activities and consulting services in Vietnam today by number of years of service

		Under 10 years		10–15 years		16 years or over	
		Quantity	%	Quantity	%	Quantity	%
Management of related S&T institutes	Poor	0	0	0	0	0	0
	Average	15	78.9	20	60.6	7	100.0
	Good	4	21.1	14	39.4	0	0
Consulting services	Poor	14	73.7	30	88.2	7	100.0
	Average	5	26.3	4	11.7	0	0
	Good	0	0	0	0	0	0

Source: Survey data from the project KX06.06/11-15 (Dao Thanh Truong, 2016a)

good level and 26.3% at the average level. The group working 10 to 15 years also has an assessment of this criterion, with 61.5% at the good level, 29.4% at the average level, and 9.1% at the weak level.

The weak level is selected most by the group "16 years or more", at 71.4%. The "good" level is the most popular choice among the working group "less than 10 years", at 73.7%. This proves that the group working for less than 10 years—the manpower group younger than the other two groups—has a more positive look.

For the criteria for the management of institutes relating to S&T activities in Vietnam (Table 3.9), all the three groups rated over 60% for the average assessment. The group "16 years or more" had the highest rate with 100% evaluating at the average level. The group "10–15 years" had the lowest rate with 60.6%. None of the three groups evaluated this criterion as weak. The "good" level has the highest rate of selection at 39.4% by the group "10–15 years", the group "under 10 years" had the rate of 21.1%, while the group "16 years or more" evaluated at 0%.

With the evaluation criteria for consulting services, there were similarities between the groups. All the three groups rated above 70% for the "weak" level. The group "16 years and over" rated 100% as weak, the group "10–15 years" rated 88.2% and finally the group "under 10 years" rated 73.7%. It can be seen that as the number of working years increases, the rating of consultancy services as weak also increased. With the average level, the rating was below 30% and no group rated the consulting service at a good level.

Table 3.10 Assessments of S&T support and cooperation and technology development from institutes and from other universities by number of years of service

		Under 10 years		10–15 years		16 years or over	
		Quantity	%	Quantity	Quantity	%	Quantity
S&T support and cooperation, technology development from other institutes	Poor	4	21.1	12	35.3	0	0
	Average	15	78.9	22	64.7	7	100.0
	Good	0	0	0	0	0	0
S&T support and cooperation, technology development from other universities	Poor	5	26.3	9	26.5	0	0
	Average	10	52.6	25	73.5	0	0
	Good	4	21.1	0	0	7	100.0

Source: Survey data from the project KX06.06/11-15 (Dao Thanh Truong, 2016a)

All the groups rated over 60% for an average evaluation of the criteria "S&T support and cooperation, industrial development from national/local research institutes". The group "16 years and over" rated 100% as average, the group "less than 10 years" with 78.9%, and finally the group of "10–15 years" with 64.7%. For the weak level, the group "10–15 years" rated the highest at 35.3%, the group "under 10 years" rated 21.1%, and the group of "16 years or more" rated 0%. None of the three groups evaluated at the good level.

In terms of S&T support and cooperation as well as industrial development from universities, there were differences in the views of the groups of less than 16 years and 16 years or more. For the group "16 years or more", 100% evaluated the criterion at a good level. However, in the group "less than 10 years", only 21.1% rated at this level, and 0% for the group "10–15 years". For the average rating, both groups "under 10 years" and "10–15 years" rated 52.6% and 73.5% respectively (Table 3.10).

For the criterion of assessing business needs to innovation, over 60% of all three groups rated it as average. The group "16 years or more" rated 100% at the average level, the group "10–15 years" rated 70.6%, and 63.2% for the group of less than 10 years. Meanwhile, for the weak assessment, only the group of "under 10 years" rated as weak with 36.8%. The group "10–15 years" rated the criteria as good with 29.4%.

In the assessment of the legal environment, it is easy to see a clear difference among the groups and the evaluation levels. For the weak level,

Table 3.11 Assessments of business needs to innovation activity and regulatory environment by number of years of service

		Under 10 years		10–15 years		16 years and above	
		Quantity	%	Quantity	%	Quantity	%
Business needs to	Poor	7	36.8	0	0	0	0
innovation activities	Average	12	63.2	23	70.6	7	100.0
	Good	0	0	10	29.4	0	0
Regulatory	Poor	4	21.1	3	9.1	4	57.1
environment	Average	10	52.6	15	45.5	3	43.9
	Good	5	26.3	15	45.5	0	0

Source: Survey data from the project KX06.06/11-15 (Dao Thanh Truong, 2016a)

Table 3.12 Assessment of intellectual property protection, telecommunications service quality, and IT support for innovation activities by the number of years of service

		Under 10 years		10–15 years		16 years or over	
		Quantity	%	Quantity	Quantity	%	Quantity
Protection of	Poor	4	21.1	5	15.1	0	0
intellectual property	Average	15	78.9	13	39.4	0	0
	Good	0	0	15	45.5	7	100.0
Quality of	Poor	4	21.1	3	9.1	0	0
telecommunication	Average	10	55.6	10	30.3	0	0
service and IT	Good	5	26.3	20	39.4	7	100.0
support for							
innovation activities							

Source: Survey data from the project KX06.06/11-15 (Dao Thanh Truong, 2016a)

more than 50% (57.1%) of the group "16 years and over" rated as such, while the group "10–15 years" had only 9.1%. The groups all rated over 40% for the average level, with the group "under 10 years" rating the highest at 52.6%. For the good level, the group "10–15 years" gave the highest rating at 45.5%, while the group "from 16 years" gave the lowest rate of 0% (Table 3.11).

For the criterion of intellectual property protection, there were also differences in the views of the three groups. While the group "16 years or more" rated 100% as good, the group "under 10 years" rated at 0%. The

Table 3.13 Assessment of the financial resources' ability to support innovation activities by the number of years of service

| | | Under 10 years | | 10–15 years | | 16 years or more | |
		Quantity	%	Quantity	Quantity	%	Quantity
Financial resources' ability to support innovation activities	Poor	13	68.4	7	14.3	0	0
	Average	0	0	20	66.7	0	0
	Good	6	31.6	7	14.3	7	100.0

Source: Survey data from the project KX06.06/11-15 (Dao Thanh Truong, 2016a)

views between these two groups were completely different. For the average level, the group "less than 10 years" had the highest rate of 78.9% and it also gave the highest rate of weak assessment at 21.1%.

In terms of the quality of telecommunications service and IT support for innovation, the groups had fairly positive ratings at average and good levels. With the weak level, the highest rate was 21.1% in the group "less than 10 years". The group "16 years or more" had the lowest rate at 0%. With the average level, the group of "under 10 years" had the highest rate with 55.6%. Hundred percent of the group "16 years or more" rated this criterion as good. It can be seen that the group of young workers had a less positive assessment than the groups who worked longer (Table 3.12).

In terms of the financial ability to support innovation activities, there were clear differences. While the majority of the group "less than 10 years" rated this criterion as weak with 68.4%, the "10–15 years" group rated it highest as average at 66.7%, and the group "16 years or more" rated 100% at a good level (Table 3.13).

- *Some results of STI activities in research institutes*

Along with the trend of integration, the quantity and quality of international scientific publications have become an important measure, with the objective index reflecting the development of S&T and the scientific performance of each country. For the scientists, international publications bring personal benefits such as providing evidence of research achievements, creating professional cooperation opportunities, and promoting international integration. There is also an obligation to share, contribute to human knowledge, and enhance the presence of the country's science.

Table 3.14 Number of research projects conducted by research institutes, 2009–2013 (by average)

	2009	2010	2011	2012	2013
State level	5.25	8	3.5	3	4.1
Ministerial level	3.9	6.25	8.25	8.5	9.6
Provincial level	0.5	3.75	0.4	1	1.2
Grassroots level	11.25	4	7.75	9	9.6
Other topics (international cooperation)	2.9	1	2.12	2.1	2.1
Average total	23.8	23	22.02	23.6	26.6

Source: Survey data from the project KX06.06/11-15 (Dao Thanh Truong, 2016a)

International scientific publication often refers to articles published in scientific journals and books, as well as internationally recognized inventions.

From 2009 to 2011, there was an average reduction of 1.78 topics in a research institute, equivalent to 8.08%. By 2012, however, the number of topics in the research institutes increased slightly, averaging 1.58 topics per institute equivalent to 7.18%. The number of topics in institutes received continued to increase in 2013 with an average increase of three topics/institute, or 12.7%.

In terms of types of topics, the state-level project had a big change. From 2009 to 2010, the number of topics of presiding institutes increased on average by 2.75 topics/institute. However, from 2011 to 2012, there was a slight decrease in the number of topics—until 2013 when it increased again, but not as many as in 2009 to 2010, with 2.33 topics/institute/year. The numbers continued to increase in the following year, but not by as many as this period.

For ministry-level topics, the number of ministerial-level projects presided by research institutes increased year by year but increased the most in the period 2009 to 2010, with an increase of 2.35 topics/institute. In the following years, the average number of topics of a presiding institute increased but not as many as in 2009 to 2010.

For provincial-level topics, the period 2009 to 2010 still saw the strongest growth, with an increase of 3.25 topics/institute. After that, in 2011, there was a sharp decline to 0.4 topics/institute (reduction of 3.35 topics). The following years maintained a slight increase.

There were not many grassroots-level topics in 2010. From 2009 to 2010, the number of grassroots-level topics decreased by 2.8 times (from

Table 3.15 Number of articles, books, and IP objects of research institutes, 2009–2013 (by average value)

	2009	2010	2011	2012	2013
International specialized journal	7.8	10.5	11.6	15.5	15.5
Domestic specialized journal	9.4	12.2	16.3	19.3	30.8
Domestic published monograph book	3.8	2.6	3.3	3.5	3.4
Other results	3.3	0	0	0	0
Average total	26	25.3	31.4	38.5	50.1

Source: Survey data from the project KX06.06/11-15 (Dao Thanh Truong, 2016a)

11.25 to 4 topics/institute). The following years saw a growth in the number of topics, with an average increase of 1.4 times (Table 3.14).

The number of S&T publications in S&T journals, especially prestigious international S&T journals, is an indicator used by many countries in the evaluation of S&T productivity of a country or territory. From 2009 to 2010, on average, there were 34.26 articles, books, and IP objects posted at each institute. Compared with other countries in the world, in terms of the quantity of S&T publications, Vietnam was ranked 62nd in the world, after Thailand by 19 grades and after Malaysia by 20 grades, but ranked ahead of Indonesia by 2 grades and the Philippines by 7 grades.

In terms of the total number of articles, books, and IP objects, it is found that there was a slight decrease of 2.8% from 2009 to 2010. Then there was a gradual increase in the following years with an average growth rate of 6.2%. Particularly from 2012 to 2013, the number of articles published increased 11.6 articles/institute, or 30.1%.

For international specialized journals—an important criteria for evaluating S&T productivity—the number of articles increased gradually in the period 2009 to 2012 with an average rate of 1.9% and leveled off in 2013.

The number of articles published in domestic specialized journals increased gradually over the years, with an average increase rate of 4.28%, from 9.4 to 30.8 articles/institute.

The number of research papers and inventions granted by institutes was quite modest compared to the regional countries aforementioned. This modest ranking is also consistent with the number of patents registered in the United States and the creative index ranked by the World Intellectual Property Organization (Table 3.15).

The main reasons for this include: inadequate budget allocation for research; barriers to the English language; inadequate awareness about the importance of international publications; inexperienced and 'unpopular' publications; lack of remuneration policies to encourage scientists to publish internationally; very few domestic scientific journals in English; and no standards available for assessing scientific effectiveness in accordance with international standards where international publication is used as an objective measure. However, there has been no significant solution to improving the situation so far.

In terms of the number of topics each individual undertakes in each institute, it is found that there was about 0.8 topics/year/institute. In the period 2009 to 2013, the year 2012 had the highest rate of 1.06 topics (Table 3.16).

According to the data, the number of articles written by experts and scientists is generally quite low, but there was a significant improvement in 2013: from 0.86 posts/person up to 1.94 articles/person/year. The number of articles, though increasing, is still slow and low. But, in general, researchers and scientists need to take new steps, and increase the productivity of S&T. Compared with organizations and individuals of industrialized countries, the R&D capacity in Vietnam is still low. Often, the quality of applications for inventions and utility solutions of Vietnamese owners is not high, mainly expressed in the poor description (not fully described, homogeneous, lack of clarity), and failing to illustrate the industrial applicability of the solution to field experts' solution appraisal. As such, the likelihood to grant a patent is not high. Many Vietnamese applicants do not know or cannot search patent information, an extremely important source of information for R&D activities. In addition, the perception of Vietnamese application owners on protection and management of industrial property rights for inventions is still limited. Some solutions are revealed publicly by the application owners for a long time before submitting the application of patent protection and utility solution, which makes

Table 3.16 Average number of topics of researchers from research institutes, 2009–2013 (by the average value)

	2009	2010	2011	2012	2013
State level	0.02	0.10	0.03	0.23	0.23
Ministerial level	0.10	0.10	0.23	0.20	0.20
Provincial level	0.03	0.03	0.03	0.03	0.20
School level	0.33	0.23	0.30	0.17	0.13
Other topics	0.17	0.37	0.10	0.43	0.23
Average total	0.65	0.83	0.69	1.06	0.99

Source: Survey data from the project KX06.06/11-15 (Dao Thanh Truong, 2016a)

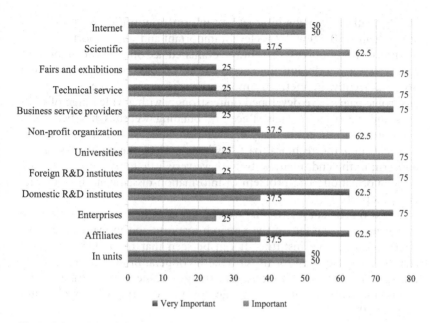

Chart 3.11 Importance of information sources with R&D and innovation activities at research institutes (in percentage). (Source: Survey data from the project KX06.06/11-15)

the solution stated in the application lose its novelty and be no longer protectable.

As the world transforms into a 'global information society', each country requires the exploitation and use of information as a fundamental and important resource for national and S&T development.

Chart 3.11 shows two important levels of assessment, which are explained as follows. The survey gives three levels: very important, important, and unimportant. However, the survey data shows only results at two levels: important and very important. The unimportant level is therefore not included in the table.

According to the data, information sources from business service providers and information from enterprises are two sources of information that are considered to be very important for R&D activities in research institutes, having the same rate of 75%. Meanwhile, sources of information from foreign R&D institutes, universities, technical service providers, and exhibition fairs have the most important rating of the same 75%. Sources of information from the internet and information in the institutes/units are evaluated 50–50 with two levels: important and very important.

Information sources are always evaluated as one of the important resources for developing the country in terms of economy, society, and S&T. Currently, in Vietnam, information resources on the website and electronic pages have developed quite strongly with a wide range of websites and electronic portals. Recently, building websites has become a widespread movement in many organizations and institutions. But, unfortunately, after the technical part has been solved, the content is of less concern; in the end, many websites have very little impact due to having poor information. The problem of information dissemination is a large, complex problem.

Based on the survey results, we can see that there is a big difference between the viewpoint of the research institute and individuals operating/working at the institute. It can be seen that the individual group assesses a close degree for the criteria of cooperation with foreign R&D units, non-profit organizations, and the government. The organization group mostly rates at the average for the criteria, with 4/7 criteria over 60%.

For the criterion of cooperation with the host organization and other research institutes, organizations assessed it as mostly average at 42.5%. Individuals rate it the highest at 66.6%.

For the domestic R&D unit cooperation, 75% of organizations and 58.8% of individuals rated this criterion as average. The close degree is evaluated more highly by individuals than that of organization at 29.4% compared with 12.5%. The weak level is rated more by organizations at 12.5%, compared to 11.8% by individuals.

In terms of foreign R&D unit cooperation, there is a big difference between the evaluations of the two groups. While the organization group rates 75% as average, the individual group rates approximately 70.6%. For the criterion of cooperation with non-profit organizations, the organization group rates it as mostly average (62.5%) and the individual group thinks that the two sides cooperate weakly (87.6%).

For cooperation with businesses, both groups have the highest average rating (35% for the organization group and 47% for individuals). However, in the organization group, the rating is weak and average at 32.5% (Charts 3.12 and 3.13).

In the organization group, four out of the seven evaluation criteria are rated as average with more than 50%. The criterion of cooperation with

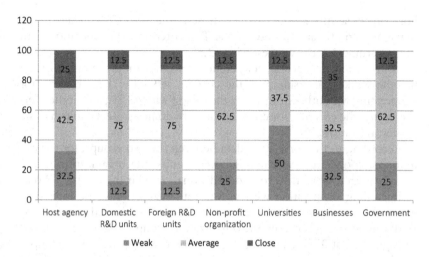

Chart 3.12 Evaluation of institutes on the level of cooperation of units in S&T service activities and innovation (percentage). (Source: Survey data from the project KX06.06/11-15)

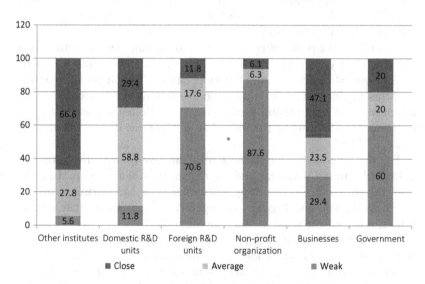

Chart 3.13 Evaluation of individuals on the level of cooperation of units in S&T service activities and innovation (percentage). (Source: Survey data from the project KX06.06/11-15)

host organizations is rated as the closest one with 75%, whereas the remaining criteria are all under 38%. The criterion for cooperation with the government is the weakest with 52.5%.

For the individual group, cooperation with the non-profit organization has the weakest assessment of the criteria, at 94.1%. The criterion for cooperating with other institutes is the closest with 83.3%, and at the average rating, the criterion for cooperation with domestic R&D units has the highest rate at 64.7 %

According to the correlation data between the two groups, for the criterion of cooperation with foreign R&D units, the organization group rated it at 75%, while the group of individuals rated it the most at 58.6%.

Regarding cooperation with non-profit organizations, the individual group does not have an average assessment, but 94.1% rated this as weak. With this criterion, however, the organization group assesses the weak and average levels at 37.5%. The importance level of the organization group is much higher than that of the individual group, at 25% and 5.9% respectively.

The criterion of cooperation with businesses also differs between the two groups. While the organization group rated mainly average at 50%, the group of individuals only rated the same at 11.8%. The individual group has a higher close rating (47%) compared to the remaining two ratings, but it is still below 50%.

In terms of the relationship in technology development activities, with the individual group, five out of the seven evaluation criteria are rated weak at over 50%. The criterion for cooperating with non-profit organizations has the highest weak rating at 86.7%. In terms of cooperation with businesses, the organization group rates it as the highest level of close cooperation (55%), while the individual group rates it as moderate and weak (37.5%).

For the criterion of cooperating with the government, the group of individuals rates 60% as weak and the remaining 40% as average. In addition, the criterion of cooperating with universities and organizations are 62.5% weak. The link between management organization (state), institute, and school improved, but not significantly.

For the correlation between two groups of individuals and organizations with three criteria—cooperation of domestic R&D units, foreign R&D units, and non-profit organizations—the group of individuals rates it as weak cooperation at over 60%. Meanwhile, the organization group rates the cooperation at an average level with a rate of over 50% (Charts 3.14 and 3.15).

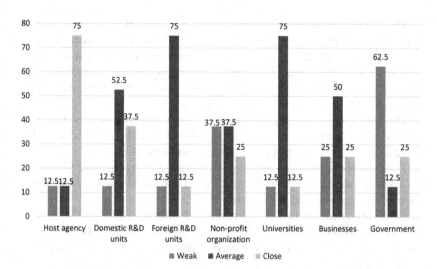

Chart 3.14 Evaluation of institutes on the level of cooperation of units in scientific research activities (percentage). (Source: Survey data from the project KX06.06/11-15)

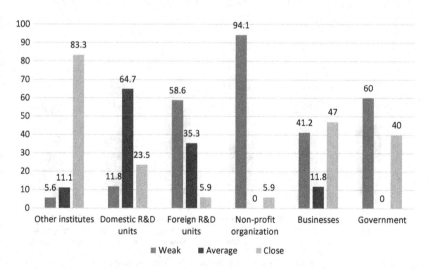

Chart 3.15 Evaluation of individuals on the level of cooperation of units in scientific research (percentage). (Source: Survey data from the project KX06.06/11-15)

The survey explores the main reason why research institutes work with the aforementioned partners in scientific research and industrial development activities in the organization. Individuals assessed the reasons for establishing long-term strategies as very important at 57.1%. The transfer of production know-how is second most important at 52.5%.

The reason of shortening the time to access the market is rated by individuals as the least important (23.8%), followed by the demand for STI (14.3%), and the transfer of production know-how (4.8%).

The cost and cause of shortening the market access time have the highest average rate of 52.4%, while the reason for establishing a long-term partner and the demand for STI was 38.1%.

In investigating the causes and realities of linkages and R&D activities, the survey explored the viewpoints of individuals to promote the development of the S&T system as well as the factors that innovation requires (Chart 3.16).

According to survey results, financial factors for enterprises had the highest percentage (100%) and are assessed as necessary to promote the STI system. Ranked second is the issue of cooperation relation with businesses (95%). The factor of creating an environment to promote scientific

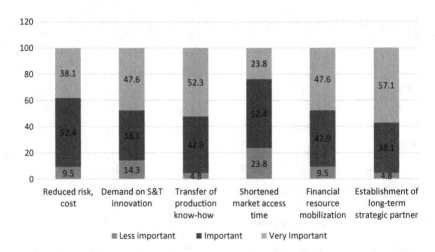

Chart 3.16 Reasons for cooperation between research institutes and partners. (Source: Survey data from the project KX06.06/11-15)

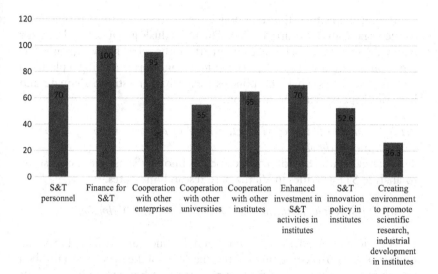

Chart 3.17 Factors in the personal point of view needed to promote the development of STI system (percentage). (Source: Survey data from the project KX06.06/11-15)

research and industrial development in research institutes has the lowest rate of 26.3%. The remaining factors have a ratio of rating as necessary, with more than 50% (Chart 3.17).

Thus, financial issues are always the most important issue to develop the country and the S&T system. In addition, human resources, policies, and relationships are also necessary and sufficient factors to complete and develop the S&T system.

In terms of the degree of difficulty, there is a difference between the perspective of the research institutes (organization group) and the perspective of individuals working at the institution (individual group). In the organization group, 88.3% say that institutes face no difficulty, while the individual group thinks that some difficulties still exist (50%).

With regard to the local policy and university capacity, both groups assess that some difficulties still exist. Particularly in the area of local policy, the individual group assessed it to be most difficult (71.6%) compared to other levels and factors. The organization group rates this factor less difficult at 45%.

Among the factors, the state policy seems to be the most problematic to the organization group (46.7%). The individuals group rated the factor below 30% with a difficult assessment. It can be seen that Vietnam's current policies are not really the driving force, and it is launching a platform for S&T activities when the policies themselves are still inadequate and have not kept up with the situation.

3.1.3.2 Current Status of STI Activities of Vietnam Academy of Science and Technology

The Vietnam Academy of Science and Technology is used as a case study to analyze STI activities in Vietnamese research institutes.

(a) *STI System of Vietnam Academy of Science and Technology*

The Vietnam Academy of Science and Technology has 51 subordinate units including 6 assisting units of the chairman of the institute established by the prime minister, 34 scientific research non-business units (27 units established by the prime minister and seven units established by the chairman of the institute), 6 other non-business units (5 units established by the prime minister and 1 unit established by the chairman of the institute), 4 self-financing units, and 1 state-owned enterprise.

The units of the institute are located in Hanoi, Ho Chi Minh City, and some other provinces. In addition, the institute also has a system of over 100 camp stations belonging to 17 specialized research institutes that are distributed in 35 provinces and cities that characterize almost all geographical regions of Vietnam (plain, coastal, midland, mountainous, and island areas) for survey, investigation, data collection, experimental deployment of geology, geography, geodynamics, geography, environment, resources, and material testing.

For many years, the Vietnam Academy of Science and Technology has four key national laboratories on gene technology, network technology, electronic materials and components, and plant cell technology, along with many other research institutes at the institutional level. The institute has pilot production areas to directly serve the technology development work, bringing scientific research results into practice.

(b) *S&T Human Resources of Vietnam Academy of Science and Technology*

In general, the institute's organizational structure is quite stable, complete, and synchronous. The workforce has a high level of knowledge with many experts and is evenly distributed in most areas of natural science. High-quality human resources have always been the institute's strength (compared with R&D units and universities in the country) (Chart 3.18).

- As of September 2017, the Vietnam Academy of Science and Technology has a total of over 4000 officials and employees, including 2351 permanent staff, 45 professors, 150 associate professors, 26 doctors of science, 838 doctors, 869 master's holders, and 550 officials and employees with university degrees (Chu Thi Hoai Thu, 2018). In the S&T human resource structure of the institute, 71.9% obtained master's and doctoral degrees, while 3.5% of the human resources are professors and associate professors. However, high-quality manpower tends to be mobile or retired. Additional replacements are not enough, which causes uncertainty regarding the staff (Chart 3.18).

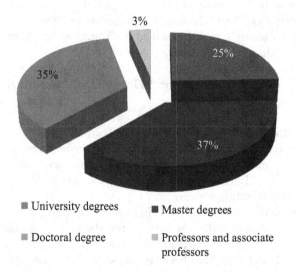

Chart 3.18   Distribution of scientific staff force in Vietnam Academy of Science and Technology in 2017. (Source: Vietnam Academy of Science and Technology, 2017)

Table 3.17 Changes in academic degrees of S&T human resources of Vietnam Academy of Science and Technology, 2012–2017

Year	Doctor of Science	Doctor	Master	University degree
2012	36	662	714	826
2013	35	706	781	794
2014	31	721	858	742
2015	45	748	915	703
2016	32	801	883	578
2017	26	838	869	550

Source: Vietnam Academy of Science and Technology, 2017

From 2012 to 2017, the number of doctoral and master's degree holders tended to increase over the years. The number of doctorates increased by 1.26 times and that of master's increased by 1.21 times (Table 3.17).

– The S&T human resources of the academy are present in all three territories—north, central, and south—but distribution of the institute's S&T human resources is not really reasonable among specialized institutes in both research directions and regions or territories. While the southern region is a vibrant economic area, the academy's highly qualified S&T staff is very thin, failing to meet development needs.

(c) Performance results of the STI system of Vietnam Academy of Science and Technology

The Vietnam Academy of Science and Technology is one of the two largest research institutes in the country. So far, the institute has had many sub-institutes and annually carries out various activities of association and cooperation in STI. That success can be valuable lessons for other organizations in the operation process.

International cooperation on S&T: Implementing open policies and guidelines in the international relations of the party and state, the Vietnam Academy of Science and Technology has gradually strengthened and expanded previous traditional relations that have been reestablished. The academy has actively cooperated with countries that have advanced S&T, such as the United States,

Germany, Japan, Canada, and Korea. In recent years, the Vietnam Academy of Science and Technology has also signed an S&T cooperation agreement with many S&T organizations in the world including France, Japan, and Korea.

Technology transfer activities: Technology transfer activities at the Vietnam Academy of Science and Technology are based on the research results of research institutes in the academy:

- *For the results of Official Development Aid projects on the universe*: Small satellite projects observing natural resources, environment, and natural disasters such as VNREDSat-1. The package of designing, manufacturing, and launching satellites (worth 55.2 million Euro) has completed the effective assembly and integration of the VNREDSat-1 satellite. There is also the ongoing second small satellite project, VNREDSat-1B, to observe natural resources, environment, and natural disasters.
- *In the field of information technology*: Vietnamese–English, English–Vietnamese two-way translation software products on the Android operating system, ViNAS Solution; an electronic newspaper-to-voice conversion system; and the building of a 3D human body model for research and teaching in medicine.
- *In the field of physics, mechanics, and materials science*: Scientists of the Institute of Mechanics have studied and developed computational software. About 60 new-generation machines have been installed and transferred to gold and silver trading and processing enterprises in 20 provinces and cities nationwide. A team of researchers has developed a device that allows automatic treatment of blood purifier after treatment for reuse for artificial-kidney patients.
- *In the fields of chemistry and the environment*: There are many achievements including Technological process of steel production from red mud with a scale of 10 tons of red mud/batch and the efficiency of iron recovery reaching over 70%; technology and equipments for extracting oil from Aquilaria trees by industrial-scale CO_2 method; deploying the production of functional food products (such as Fucogastro, Fucoidan FucoAntiK and Fucoidan FucoUmi), which have a number of medical benefits such as supporting the immunity to improve resistance to cancer and supporting cancer treatment; completing the technology of manufacturing low-cost bandages that contain silver nano-particles to treat hard-to-heal wounds on humans.

- *In the fields of biology and ecology*: R&D of bio-diagnostic kits for influenza A/H5N1; R&D of a new rice variety, 'Nam Dinh 5'; and a 200-liter pilot seaweed biomass fermentation system with portable design in 40-foot container. In 2012, the Academy's ecologists published 62 new species for world science.
- *In the field of earth science*: On-time land saving in the area of Song Tranh 2 hydropower plant; conducting geophysical surveys in the Ninh Thuan area to provide a scientific basis for the evaluation of survey documents on the selection of construction sites for the Ninh Thuan 1 and Ninh Thuan 2 nuclear power plants with Japanese and Russian consultants.
- *Regarding the publication of IP works and registration*: In 2012, the institute announced a total of 1698 scientific projects. Particularly, there were 401 publications in prestigious international journals meeting ISI (SCI and SCI-E) standards. This was an increase of 20% from 2011 (334), including many works published on high-impact-ratio journals. In 2012, scientists of the institute published 64 monographs with an increase of 88% from 2011 (34). In 2013, the Vietnam Academy of Science and Technology announced a total of 2298 scientific projects, an increase of 36% compared to 2012. There was a total of 660 international articles published in 2013, an increase of 11% compared to 2012. There were 435 international publications in prestigious international journals meeting the ISI (SCI and SCI-E) standards, an increase of 0.9% compared to 2012, many of which were published in high-impact-ratio journals.

(d) *Finance for S&T activities*

The Vietnam Academy of Science and Technology effectively implemented the Budget Law and the Law on S&T to improve scientific management as well as enhance the responsibility of the host unit and the chief author when performing tasks. The topics at the end of the implementation period are checked and evaluated. Products of the topic after acceptance can be transferred to practice or have orientations for the next development steps.

From 2013 to 2018, the total allocated budget of the academy increased by 3.2 times from VND784 billion in 2013 to VND2535 billion in 2018 (Charts 3.19).

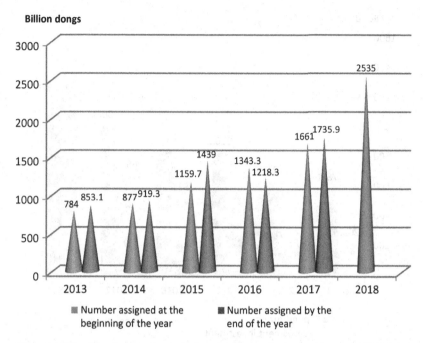

Billion dongs

■ Number assigned at the beginning of the year   ■ Number assigned by the end of the year

Chart 3.19 Total annual operating budget of the academy, 2013–2018 (excluding foreign capital). (Source: Vietnam Academy of Science and Technology, 2017)

In the annual operating budget structure of the academy, funding for development investment increased from VND205.6 billion in 2013 to VND1356.8 billion in 2018 (Chart 3.20).

Regarding the academy's organizational trend, besides the member units, the Vietnam Academy of Science and Technology also coordinates closely with other ministries to implement programs to enhance research capacity as well as formulate strong research institutions and basic research groups in accordance with the objectives of Decision No. 2133/QD-TTg dated 1 December 2011 on "Vietnam Science and Technology to 2020 and orientation to 2030". There are 35 specialized research institutes, subordinate non-business organizations, one Academy of Science and Technology, and 15 S&T enterprises (spin-off enterprises). They have 3500 staff and 1700 contract workers, half of whom are scientific staff with doctoral and master's degrees. The focus is to achieve a rate of

Billion dongs

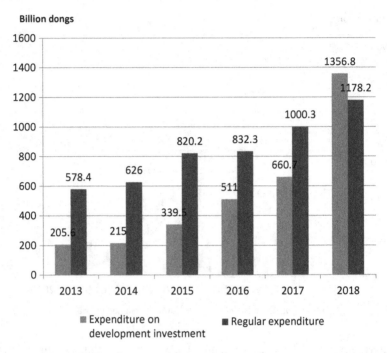

Chart 3.20 Structure of the annual operation budget of the academy in the period 2013–2018 (excluding foreign capital; budget allocated at the beginning of the year). (Source: Vietnam Academy of Science and Technology, 2017)

researchers/junior assisting researchers that would create a reasonable operational structure of specialized institutes. The number of internationally published scientific works and the number of patents registered for protection has also doubled.

3.1.4 *STI System of Enterprises in the Trend of International S&T Integration*

It is undeniable that investment in S&T has helped many businesses improve product quality and increase market competitiveness. For S&T activities at universities or research institutes, it is encouraging to publish the result of a research project in a scientific journal, but for businesses, this is not a priority. What they are interested in is understanding the market, applying S&T to increase product quality, deploying new products, meeting market demands, reducing production costs, and increasing competitiveness. The aim of the business is profit. Similarly, R&D organizations must create scientific research results according to enterprises' orders or meet market demands, which enterprises aim for. Enterprises will have to select and analyze potential value and benefits, and balance costs for scientific research results from R&D organizations. The role of the state is to provide an appropriate orientation policy for commercializing scientific research results or for funding R&D for enterprises.

Businesses in countries such as Singapore, Japan, and Korea have taken advantage of the power of S&T as the key to their growth. In Vietnam, the state also creates preferential policies for businesses investing in S&T development, such as ordering businesses to conduct scientific research to create products that serve society practically, encouraging enterprises to deduct annual revenue to invest in S&T development, and importing source technology and high technology. But with what has been achieved, Vietnam's S&T has not really brought about efficiency and growth for businesses. Vietnamese enterprises have not taken advantage of the benefits and potentials of the country's S&T. Looking at the whole STI system of Vietnam, the imprint of the enterprise is fuzzy in its association with the institute, the university, and the state.

3.1.4.1 Current Situation of S&T Human Resources of Enterprises

In the context of economic integration, when competition is growing fiercer, human resources play an increasingly important role. As Vietnam transforms from an economy based on comparative advantages by relying on cheap labor, resources, and environment, to one with competitive advantages based on promoting human resources, reliance on high-quality S&T human resources is necessary.

The results from the survey of KX06.06/11-15 (out of a total of 208 sheets, 104 were for businesses and 104 sheets were for individuals working at these enterprises) show that the research and technical personnel in enterprises, including large-scale enterprises, account for less than 40%. For small-scale enterprises, S&T personnel have the highest percentage of 37.1%. Support staff account for 17.4%, while other human resources account for 4.4%. For medium-size enterprises, the highest proportion of technical manpower is 36.4%, followed by assisting human resources with 29.5%, and other manpower is still the lowest with 6.8%. Large enterprises have a structure different from the other two enterprises. In large-scale

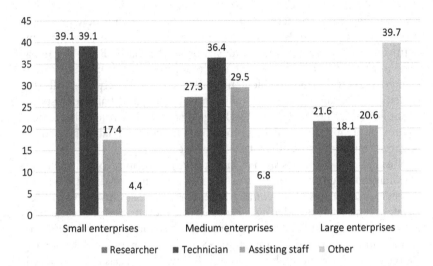

Chart 3.21 Percentage of personnel under tasks in S&T activities of enterprises classified by size of enterprises. (Source: Survey data from the project KX06.06/11-15)

enterprises, other human resources account for the highest proportion with 39.7%, assisting human resources and research human resources account for less than 25%, with technical personnel having the lowest rate at 18.1% (Chart 3.21).

Thus, we can see that research and technical manpower in enterprises accounts for a relatively low proportion, which is not commensurate with the requirements of S&T and innovation in enterprises (Table 3.18).

According to the data table, we can see that there is a clear difference between female and male manpower, as well as Vietnamese manpower and foreign manpower. There is also a difference in terms of the allocation of R&D personnel according to job functions. Human resources performing research work are much higher than technical manpower and support staff. This data is similar to the database published by the Ministry of Science and Technology in 2014. In that database, human resources comprised 14,990 (77.8%) research workers, 1423 (7.4%) technical personnel, and 1423 (7.4%) support staff.

In terms of gender relations, we can see that there are more men than women. For research human resources, female workers account for only 4.4% of the total workforce. The number of female technical workers is

Table 3.18 Average number of employees under the tasks in S&T activities of enterprises by gender (*Unit: persons*)

Manpower	Vietnamese		Vietnamese	
	Average value		*Average value*	
	Male	*Female*	*Male*	*Female*
	Mean	*Mean*	*Mean*	*Mean*
1. Research human resources (with college degree or higher), spending at least 10% of their time for R&D	14.3	0.6	0.5	0.06
2. Technical personnel (including technicians, laboratory staff with intermediate and equivalent qualifications)	12.5	1.6	0	1.1
3. Support staff (administrative and office staff)	11.2	1.6	0.09	0.4
4. Other	5.8	0.6	0.1	0.06

Source: Survey data from the project KX06.06/11-15 (Dao Thanh Truong, 2016a)

also much lower at 21.6%, and the female support staff is 17.7%. This is fully understandable because the actual data shows that the percentage of women doing R&D (57,121 people) is not high (42.4%) among all R&D personnel.

In terms of foreign human resources operating in the field of R&D in Vietnam, we can see that they account for a very low rate (4.8%), which is worrying when Vietnam is in the process of S&T integration. Besides, the number of female foreign workers working in Vietnam is also higher than the number of male workers by 2.3 times.

In terms of professional qualifications and titles of human resources operating in S&T at enterprises, there is great difference between the ratio of men and women, as well as between Vietnamese staff and foreigners.

In terms of professional qualifications, the proportion of human resources with college degrees and above is under 56%. Among these, the lowest rate of manpower with the title of associate professor in enterprises is under 6%, followed by doctoral degree at below 12%, and university degree with the highest rate of over 55% (Chart 3.22).

Among the three types of enterprises, medium-size enterprises have the highest proportion of manpower with master's degree at 42.9% and the highest associate professor rate at 5.7%.

In terms of professional qualifications and titles of human resources operating in S&T at enterprises, there are a lot of differences between the ratio of men and women, Vietnamese and foreigners (Table 3.19).

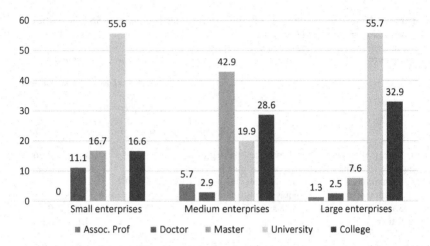

Chart 3.22 Proportion of personnel according to professional qualifications and titles by enterprise size (percentage). (Source: Survey data from the project KX06.06/11-15)

Table 3.19 Average number of employees according to professional qualifications and titles in S&T activities of enterprises (*Unit: persons*)

Nhân lực	*Vietnamese*		*Foreigner*	
	Average value quantity		*Average value*	
	Male	*Female*	*Male*	*Female*
	Mean	*Mean*	*Mean*	*Mean*
1. Divided by professional level				
Doctor	0.72	0.15	0.03	0
Master	1.317	0.49	0.08	0.03
University degree	13.25	7.1	0	0.07
College degree	1.63	0.94	0	0
2. Divided by academic title				
3. Professor	0	0	0	0
4. Associate Professor	0.3	0	0.3	0

Source: Survey data from the project KX06.06/11-15 (Dao Thanh Truong, 2016a)

According to the table, in terms of professional qualifications, those with university degrees have the highest rate (79.2%) in all S&T human resources of enterprises. On average, 16.94 people in an enterprise have a university degree. Second, the manpower with college degrees is 10%, equivalent to an average of 2.57 people/business. The proportion of human resources with a doctoral degree is the lowest at 3.4%, the average of 0.87 people/enterprise. This means that there are enterprises that do not have anyone with a doctorate degree. Thus, it can be seen that the current level of human resources in enterprises is not high and only focused on university and college levels.

Regarding the correlation between male and female human resources, the proportion of female workers is much lower than that of men. At the university level, the average number of male employees is 1.85 times higher than the number of female employees. Even in the survey data of the Ministry of Science and Technology, it is easy to see that the number of women in S&T activities and innovations of enterprises according to professional qualifications accounts for a very low rate (37.1%). The number of people who hold a doctoral degree accounts for 0.59% of all S&T human resources, 2.02% with master's degrees, 30.9% with university degrees, and 3.6% with college degrees.

The number of foreigners working in S&T fields in Vietnamese enterprises is also very low. They have a university degree or higher, and nobody has a college degree. For those who hold doctoral degrees, the majority is male, and there are 0.03 men/enterprise on average. There are more foreign workers with master's degrees—an average of 0.08 men/enterprise and 0.03 women/enterprise. For those with university degrees, there are only 0.07 female employees/enterprise.

Within the limited conditions of the scientific environment, the S&T human resources in the enterprises are still small in number and limited in capacity, failing to meet the requirements of enterprises' STI. The distribution of manpower and the qualification structure is not reasonable according to the size of enterprises. There is a shortage of good leading manpower, especially a lack of younger S&T human resources with high qualifications. Most S&T human resources are currently working in the state sector (research institutes, universities, and S&T management organizations), and this rate is still very low in the private sector and enterprises. The link between research, exchange, and cooperation in S&T research projects among enterprises and other S&T units is still limited. Only a few enterprises attach importance to linking with universities and research institutes of the same specialized fields to attract and recruit S&T workers. These include FECON Mining Joint Stock Company which is associated with the Institute of Construction Materials, Hanoi University of Civil Engineering; or Rang Dong Light Sources and Vacuum Flask Joint Stock Company, which is associated with the International Training Institute for Materials Science and the Advanced Institute of Science and Technology (Hanoi University of Science and Technology).

3.1.4.2   Financial Investment of Enterprises for R&D Activities

Every year, Vietnam invests about 2% of its total budget in S&T, which is equivalent to 0.5% of the GDP, but most are distributed to research institutes, companies, and enterprises managed by the state. Joint-stock companies (state capital less than 50%) and private enterprises are not provided with research funding and new technology application.

According to the Ministry of Science and Technology data published in the book, *Science and Technology of Vietnam in 2013*, Vietnamese enterprises spent VND1377.01 billion for R&D activities, the second highest after research institutes and centers (VND2311.02 billion) (Chart 3.23).

For the level of expenditure for each type of R&D activities (basic research, applied research, and experimental implementation), enterprises have different spending levels. There are 31.4% of enterprises that spend

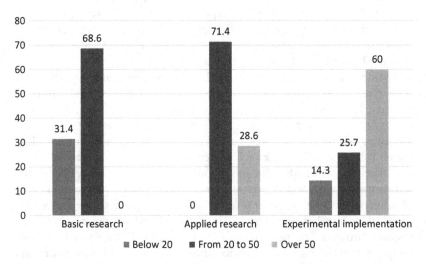

Chart 3.23 Estimated cost of enterprises for scientific research and technology development according to the above section in 2013 (percentage). (Source: Survey data from the project KX06.06/11-15)

less than 20% of the total expenditure on basic research. Most enterprises (about 68.6%) spend 20% to 50% on basic research, and none of them spends more than 50% of the budget on this type of activity.

Enterprises spend more on basic research than on applied research. Among the enterprises participating in the survey, 71.4% of them spend 20% to 50% of their budget on applied research, and 28.6% of enterprises spend over 50% on applied research.

About 60% of enterprises spend over 50% on experimental implementation. This is different from research institutes, which mostly focus their budget on applied and basic research.

According to the General Statistics Office, over 95% of the small and super-small enterprises have limited resources, especially capital, so the majority of these enterprises focus on technology innovation rather than R&D activities.

Following the research results of the Vietnam Chamber of Commerce and Industry (VCCI) experts from 2007 to 2010, which considered the total investment for S&T activities, the proportion of investment in R&D activities decreased sharply from 55.3% in 2007 to only 38.35% in 2010. Meanwhile, the average cost of each enterprise for technological innovation was constantly increasing in both value and proportion. If in 2007, on average, an enterprise only invested about VND712 million, accounting for 33% of the total investment for S&T, then by 2010, the cost of technology innovation had increased three times, reaching over VND2 billion, accounting for nearly 50% of the total investment for S&T in enterprises. It is noteworthy that enterprises mainly used their own capital to invest in S&T activities.

The total investment capital for S&T at enterprises not only accounted for a large proportion but also tended to increase, from 75.89% in 2007 to 86.06% in 2010. Meanwhile, the proportion of investment from the state budget in S&T activities at enterprises was decreasing, from 15.06% in 2007 to only 8.48% in 2010. It is worth noting that all of the above analysis data is based only on the data provided by the participating enterprises pertaining to the situation of investment in S&T at enterprises. This accounts for less than 0.2% of the total number of enterprises surveyed. (In 2007, the number of enterprises responded to S&T/Total surveyed enterprises were 271/155.771 and in 2010, it was 509/290,767). Thus, about 99.8% of surveyed enterprises are not interested in S&T activities or are not capable enough to innovate technology.

Businesses in Vietnam can invest in new or improved R&D activities based on the technology available in the market. In terms of funding dedicated to R&D activities in enterprises, the venture capital for this risky operation from the funds is very small (Chart 3.24).

According to the chart, we can see that over 80% of businesses do not receive venture capital. Small businesses have the highest rate of not receiving investment capital, at 83.3%. The large enterprise group receives the most capital among the three groups of enterprises, at 18.5%.

The number of investment funds currently operating in Vietnam is quite large, with about 56 funds. The various funds are as follows: three funds for high-tech investment; three funds for venture investment (Mekong Capital, IDGVV—IDG Venture VietNam, VinaCapital); seven funds for investment in real estate; and 43 funds for investment in listed stocks, joint-stock companies, and opportunistic investments. Even though there are many investment funds operating in Vietnam, the funds of venture capital are of a very low proportion (three out of 56 funds). It must be acknowledged that most of the venture capital in the country has not really operated with its inherent functions.

It is found that enterprises spend about 10% on S&T activities with their annual turnover (Tables 3.20 and 3.21).

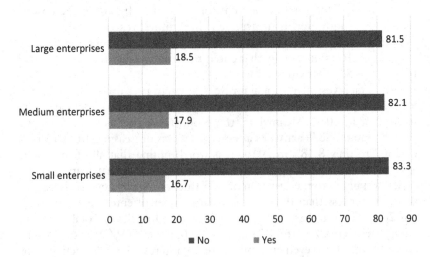

Chart 3.24 Percentage of enterprises receiving venture capital by size (percentage) (Source: Survey data from the project KX06.06/11-15)

Table 3.20 Estimated cost of S&T activities of enterprises (billion VND)

	Quantity	%
Less than 10	24	23.1
From 10 to 100	30	28.8
From 101 to 300	12	11.5
Over 300	6	5.8
Unknown/unwanted to answer	32	30.8
Total	104	100.0

Source: Survey data from the project KX06.06/11-15 (Dao Thanh Truong, 2016a)

Table 3.21 Proportion of expenditure of enterprises for S&T activities, 2013

Private capital of enterprises (percentage)	Quantity	%
Less than 20	0	0
From 21 to 50	15	14.4
From 51 to 80	15	14.4
Over 80	45	43.3
Unknown/unwanted to answer	29	27.9
Total	104	100.0

Source: Survey data from the project KX06.06/11-15 (Dao Thanh Truong, 2016a)

In the survey results, the cost of S&T activities of enterprises is mainly from private capital. Enterprises pay for this activity by themselves. In addition to private sources, enterprises seek financial resources from banks. Enterprises do not receive capital support from other sources (such as foreign enterprises, other domestic enterprises, or the government).

According to survey data, estimates of enterprises' costs for process improvement activities have a relatively low rate of below 40% for all levels. However, with an expenditure of more than 50%, this activity is much higher than the other two activities (process development and product improvement).

In terms of product development, all spending levels are below 50%. Yet, this activity has the highest proportion of enterprises with the largest budget among the three activities, at 48.6%.

Product process improvement is the only activity among the three types of activities with 60% of enterprises spending on this. In this type of activity, the spending level of under 20% accounts for 42.8%, the level between

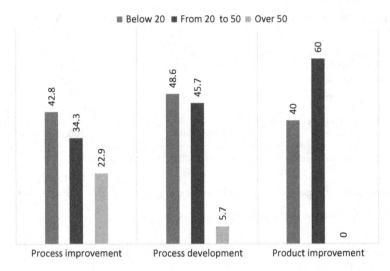

Chart 3.25 Estimated cost of enterprises for scientific research and technology development according to the above section in 2013 (percentage). (Source: Survey data from the project KX06.06/11-15)

20% and 50% accounts for 60%, and no enterprise spends over 50% of the budget (Chart 3.25).

Among the 104 enterprises participating in the survey, enterprises bearing the cost of between 20% and 50% for manpower expenditure account for the highest proportion (51.4%). Expenditure of over 50% accounts for the lowest rate, at 14.3%.

Enterprises have the highest spending rate of 20% to 50%, at 57.1%, for land and office expenses. The rate of spending below 20% has quite a good rate with 42.9%, and no enterprise spends over 50%.

The proportion of enterprises spending 20% to 50% of the budget for transport accounts for 71.5%, the highest among the three levels. Percentage of enterprises that spend less than 20% of the budget on transportation is the least, at 11.4%.

Other fees (machine repair, maintenance, etc.) comprise the highest proportion of expenditure in enterprises at 62.9%, spending between 20% and 50%. No enterprise spends more than 50% of the budget on other operations, and 37.1% of enterprises spend less than 20% of the budget on other operations (Chart 3.26).

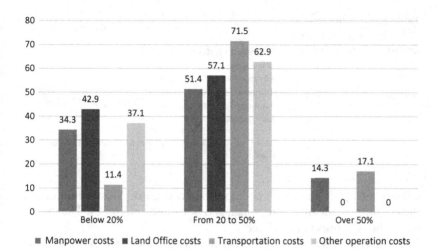

Chart 3.26 Estimated cost of enterprises for other expenses in 2013 (percentage). (Source: Survey data from the project KX06.06/11-15)

3.1.4.3 Results of STI Activities of Businesses

Currently, businesses in Vietnam also have many typical activities such as Military Telecom Corporation (Viettel), which established its own research institute in 2010 following the model of major corporations in the world. Viettel has deducted 10% of pre-tax profit for the S&T Development Fund, equivalent to VND2500 billion. With such a level of investment, only after a short time, the most important products for the information and communication technology sector of Viettel Research Institute have met the needs of the business development, and their prices are only a third of the market price.[29] A similar scenario happened with the Vietnam National Oil and Gas Group. In response to the requirement of technology development for the mining industry, in 2009, the National Oil and Gas Group cooperated with the Ministry of Science and Technology to manufacture new-generation drilling rigs for oil and gas exploitation activities. As a result, the 90-meter jack-up rig was produced. With the efforts of domestic scientists, they have designed and mastered the technology of manufacturing drilling rigs, putting Vietnam on the list of one in ten countries in the world and one of three Asian countries capable of manufacturing the 90-meter and 120-meter jack-up rigs.[30] However, the above two enterprises are large, state-owned enterprises, so the potential for

Table 3.22 Achievements of R&D activities, innovation to business activities in 2013

	Quantity	%
Improved production process efficiency	48	59.3
Given new production process	51	63.0
Improved product quality	48	59.3
Launched new products to market	63	60.6
Got patents	6	8.0

Source: Survey data from the project KX06.06/11-15 (Dao Thanh Truong, 2016a)

investment in S&T research and application is great. Most businesses in the country are medium-size, small, and non-state enterprises. The investment ability is thus not available.

Eight years after the government implemented Decree No. 80/2007 on 19 May 2007 for S&T enterprises, and seven years after Circular No. 6, which guides the implementation of Decree 80, was approved and issued by the Ministry of Science and Technology and nearly two years of implementing the prime minister's decisions on support packages for S&T enterprise development, it can be said that the results are still far from the required levels in terms of both quantity and quality.

In 2013, businesses achieved some positive results from R&D activities. According to survey data, via R&D, over 50% of enterprises achieved "improved production efficiency" (59.3%), were "given new processes" (63%), "improved quality volume of products" (59.3%), and "launched new products to market" (60.6%). Only 8% of enterprises managed to patent their results from R&D activities. The reason given by enterprises is the cumbersome registration procedure and inconsistent law (Table 3.22).

One of the factors for promoting S&T activities and maintaining S&T activities in enterprises is related to information resources. Among the sources of information that enterprises receive in scientific research and technology development activities, the source of information from domestic media partners is most used by enterprises (78.8%). Additionally, the internet and domestic information yearbook are the least accessible sources of information by enterprises (23.1%).

Even though information on the internet is extremely diverse and growing, the quality of information on websites and electronic portals is not high, so the yearbook does not really provide enough information that businesses need. On the other hand, the majority of businesses receive less than 24% of foreign S&T information. The source from fairs and exhibitions has the highest rate at 23.1%, and information from state organizations accounts for the lowest rate at 4.4%. This proves that the foreign information that businesses have is very limited. According to the survey data collected, there are opposite views between the enterprise group (organizational group) and the group of individuals who are doing S&T activities in the enterprise (individual group). The organization group believes that the source of information in the enterprise is important (60%) to very important (40%), followed by affiliates (34.3% important and 62.8% very important) and foreign supplies (22.9% important and 74.2% very important). The individual group says that information from the internet is the most important (36.1% important and 61.1% very important). On the other hand, in assessing information sources, the individual group's important and very important evaluations are quite high.

The unimportant level of the information sources is below 38%. The source of information from public research institutes has the highest percentage of individuals rating it unimportant, at 37.1%. Meanwhile, in the organization group, information from non-profit organizations has the highest unimportant assessment rate at 45.7%.

The remaining information channels have a not-very-important assessment rate of under 38%. In order to develop the innovation system and to promote S&T activities of enterprises, enterprises cooperate with some organizations and individuals.

According to survey data, there is a difference between organization group and individual group. The organization group assesses that cooperation with buyers/customers is the most important (43.8%) because businesses will know the needs and tastes of the market in order to research and deploy appropriate products. For other units and individuals of organizational groups involved in product innovation activities, cooperation with customers/buyers is considered to be the closest (65%), followed by cooperation with domestic suppliers (56.5%), and cooperation with foreign suppliers (43.7%). In addition, cooperation with other businesses has the highest average rating (55%).

The remaining organizations all have weak ratings of over 50%. Cooperation with schools has the highest rate of weak assessment (64.3%) (Charts 3.27 and 3.28).

The group of individuals primarily assesses the extent of cooperation at an average level of over 44%. In the process innovation activities, cooperation with other enterprises has the highest average rate of 69.4%. At a weak level, according to individuals, the level of cooperation with non-

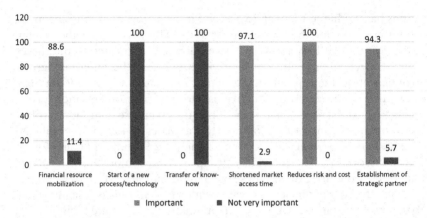

Chart 3.27 Organizations' assessment of the level of cooperation with organizations in process innovation activities (percentage). (Source: Survey data from the project KX06.06/11-15)

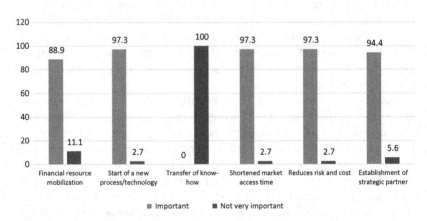

Chart 3.28 Individuals' assessment of the level of cooperation with organizations in process innovation activities (percentage). (Source: Survey data from the project KX06.06/11-15)

profit organizations has the highest rating of 50%, and the lowest rate is cooperation with buyers/customers (2.8%). Customers/buyers have the highest cooperation rating of 36.1%, and the rest are rated below 30%.

The organization group has a different viewpoint. Firstly, at a tight level, they assess that, in the innovation process, cooperation with domestic suppliers has the highest rate at 59.1%. Next, cooperation with buyers/

customers accounts for the highest proportion at 50%. The organizations assess cooperation with competitors at a weak level, with the highest rate of 61.5%.

Scientific research activities as well as processes of innovation and product innovation depend on each activity. The level of cooperation of enterprises is different. The views between individuals at businesses and organizations are different.

The survey explores the reasons why enterprises cooperate with the above partners.

The individual group working in enterprises thinks that the reason for transferring the know-how through cooperation with other organizations is not very important at 100%. For other reasons, there is an important assessment of over 85%.

In addition, from the point of view of the organization group, cooperating with organizations to start a new process and technology transfer is not very important (100%). Other reasons are considered important with a rate of over 88%. In particular, the reason for reducing risks and costs has the highest rate of 100%.

In Vietnam today, there are up to 95% of SMEs using outdated technology compared to the world, and the capacity of research and technological innovation of these enterprises is still very limited. Outdated technology and equipment seem to be one of the major causes of energy wastage, causing businesses to increase investment costs.

Among the 104 enterprises who were asked about getting advice on S&T-related policies and innovation, only 56 businesses receive consultancy, accounting for nearly 54%. Among them, 32 businesses hire consultants and 3 enterprises recruit specialized human resources, accounting for 57.1% and 5.5% respectively. There are up to 30.3%, or 17 enterprises, using self-study policy information, and 7.1% attend workshops and trainings (Table 3.23).

Table 3.23 Forms of businesses using advice on S&T-related issues and innovation

Forms of advice	Quantity	%
Participation in workshops and training	4	7.1
Self-study of information	17	30.3
Hiring consultants	32	57.1
Recruiting specialized personnel	3	5.5
Total	56	100

Source: Survey data from the project KX06.06/11-15 (Dao Thanh Truong, 2016a)

About 79% of the S&T enterprises conduct scientific research activities (Charts 3.29 and 3.30).

Majority of product or process improvement results are due to Vietnamese enterprises improving their existing technology.

In summary, Vietnamese enterprises are mainly SMEs. Their ability to self-innovate is limited, and the state has not issued effective mechanisms

Chart 3.29
Proportion of enterprises
implementing scientific
research activities.
(Source: Survey data
from the project
KX06.06/11-15)

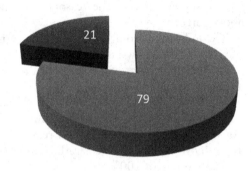

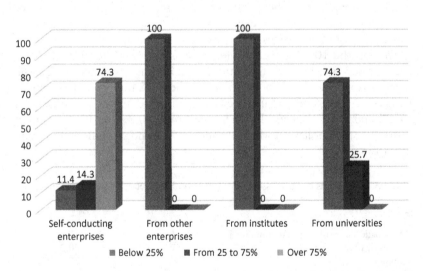

Chart 3.30 Proportion of new products/processes that enterprises implemented in 2013. (Source: Survey data from the project KX06.06/11-15)

and policies. The cohesion between research institute, university, and enterprise is still limited. The enforcement of intellectual property law in many S&T institutions is not serious, and the violation of intellectual property rights is still rampant. This limits the attraction of R&D investment in Vietnam and becomes a barrier to the formation and development of S&T enterprises in the present time.

3.2   Assessment of Vietnam STI System in the Context of International S&T Integration

Vietnam has recorded impressive achievements in economic development and poverty reduction. In the past two decades, Vietnam's GDP increased by an average of 7% a year, and the poverty rate decreased from about 60% in 1993 to nearly 12% by the end of 2011. The economic structure shifted from an agricultural base to an economy increasingly focused on industry and services. To support the development of STI, in recent years, Vietnam has made efforts to change both the mechanism and the way it works to improve the country's S&T capacity. The STI results that Vietnam has achieved show its emphasis on S&T, but Vietnam is facing competition from other countries. The global environment is becoming increasingly complex. Vietnam's general S&T level is still far behind the world's, even when compared with some leading countries in Southeast Asia. Through the surveys and analyses of the current STI situation, the author identifies the following weaknesses and limitations:

- The links between research and training, research and market, and scientists and enterprises are weak. The conversion of S&T organizations to autonomous entities has many barriers leading to the low efficiency of policies.
- Funds for S&T investment are still limited, and the use efficiency is not high. There has not been a suitable mechanism to mobilize social resources and enterprises to invest in S&T.
- The potential of S&T and the S&T workforce have been developed in quantity, but the quality has not met the requirements.
- The level of innovation in the enterprise sector is still quite low. Very few businesses set up R&D units, and most businesses do not have knowledge that can be utilized to meet the demands for innovation.
- Technical facilities and equipment are missing, yet synchronized.

- In managing S&T activities, there is a lack of appropriate priority directions, strong policies, and solutions to create breakthroughs in areas in which Vietnam has advantages.
- The expenditure norms and procedures for the actual payment of S&T tasks are inadequate. There is a lack of reliable and synchronous national statistical database on STI.

The most noticeable characteristic of Vietnam's STI system is that there is very little private sector participation. There is also low and inefficient state investment and low output in response to socioeconomic demands. In recent years, universities and research institutes have been gradually modernized. More technology enterprises have been established and developed. However, there is still a long struggle before S&T becomes a driving force for the economy.

NOTES

1. Kraemer-Mbula and Wamae (2010, p. 43).
2. Documents of the XII Congress of the Communist Party of Vietnam.
3. In pursuance to Decree No. 95/2017/ND-CP dated 16 November 2017 by the Government stipulating functions, tasks, rights, and organizational structure of the Ministry of Science and Technology.
4. Hoang Van Tuyen (2017). Technology and Innovation System in Developing Countries: Vietnam Case. *Vietnam Science and Technology Journal*, 4 April 2017, pp. 4–9.
5. Project "*Promoting Innovation Through Research, Science and Technology*" (FIRST) funded by the World Bank (WB); Project "*Vietnam—Finland Innovation Partnership Programme*" (IPP) phase 2 (2014–2018); Project "*Building Innovation Policies and Developing Business Incubators*" (BIPP) in cooperation with the Kingdom of Belgium.
6. http://ncstp.gov.vn/vi/gioi-thieu/chuc-nang-nhiem-vu.
7. Vietnam Ministry of Science and Technology (2016). *Vietnam Science and Technology White Paper 2015*, p. 67.
8. Vietnam Ministry of Science and Technology (2016). *Vietnam Science and Technology White Paper 2015*, p. 73.
9. Key laboratories are located in 7 areas: biotechnology (5 labs); information technology (3 labs); material technology (2 labs); machine manufacturing and automation technology (2 labs); petrochemical (1 lab); energy (1 lab); infrastructure (2 labs). The above key laboratories are located in 13 research institutes and 3 universities under 8 ministries and branches.

10. Vietnam Ministry of Science and Technology (2014). *Vietnam Science and Technology White Paper 2015.*
11. Vietnam Ministry of Science and Technology (2016). *Vietnam Science and Technology White Paper 2015,* p. 75.
12. Pursuant to Circular No. 03.2015/TT-BKHCN issued on 09 March 2015, regulating the Sample Charter on organization and operation of Science and Technology Development Funds of ministries, provinces, and centrally run cities.
13. Nguyen Van Anh, Khuat Duy Vinh Long and Le Vu Toan (2012). About Activities of the Local Science and Technology Development Fund. Source: https://thongtinphapluatdansu.edu.vn/2012/05/10/10-05-2012/.
14. Vietnam Ministry of Science and Technology (2014). *Vietnam Science and Technology White Paper 2013,* pp. 83–85.
15. Vietnam Ministry of Science and Technology (2016). *Vietnam Science and Technology White Paper 2015,* p. 62.
16. Signed by the President on 22 June 2000 and taking effect from 1 January 2001.
17. Vietnam Ministry of Science and Technology (2016). *Vietnam Science and Technology White Paper 2015.*
18. Vietnam Ministry of Science and Technology (2016). *Vietnam Science and Technology White Paper 2015,* p. 129.
19. Vietnam Ministry of Science and Technology (2016). *Vietnam Science and Technology White Paper 2015,* p. 132.
20. According to statistics by Scientometrics for Vietnam. Source: http://scientometrics4vn.com/category/recent-posts/vietnam-vs-asean/2018-vietnam-vs-asean/.
21. Department for Intellectual Property (2017). *Annual Intellectual Property Report 2017,* p. 94
22. Vietnam Ministry of Science and Technology (2016). *Science and Technology White Paper 2015,* p. 73.
23. Source: Ministry of Education and Training (2017). *Research Report on Science and Technology Activities in Higher Education Institutions in the Period of 2011–2016 and Proposed Solutions for Development in the Period of 2017–2025.*
24. Vietnam Ministry of Science and Technology (2016). *Science and Technology White Paper 2015,* p. 52.
25. Vietnam Ministry of Education and Training (2017, p. 21).
26. www.most.gov.vn.
27. Quacquarelli Symonds (QS) is a British company specializing in world education research founded by Nunzio Quacquarelli in 1990. From 2004 to 2009, QS collaborated with Times Higher Education (THE) to annually issue a ranking of world universities and database providers to rank.

28. Ministry of Science, Technology, Science and Technology of Vietnam 2013, Science and Technology Publishing House, 2014, pp. 207–208.
29. Viettel develops many products for defense, http://www.viettel.com.vn/.
30. Vietnam National Oil and Gas Group Information Portal, http://www.pvn.vn/.

REFERENCES

Chu Thi Hoai Thu. (2018). *Thesis on Financial Policy Regulating Social Mobility of S&T Human Resources in the Context of International Integration (Case Study of Vietnam Academy of Science and Technology)*. Hanoi: Faculty of Management Science, VNU—University of Social Sciences and Humanities Vietnam.

Dao Thanh Truong. (2016a). *Chính sách Khoa học, Công nghệ, và Đổi mới (STI) của Việt Nam trong xu thế hội nhập quốc tế: Thực trạng và giải pháp* [Science, Technology, and Innovation Policies of Vietnam in the Trend of International Integration: Situations and Solutions]. Hanoi: Thế Giới Publishers.

Dao Thanh Truong. (2016b). *Di động xã hội của nhân lực khoa học và công nghệ trong bối cảnh hội nhập quốc tế: Lý luận và thực tiễn* [Social Mobility of Science and Technology Human Resources in the Trend of International Integration: Theories and Practices]. Hanoi: Thế Giới Publishers.

General Statistics Office. (2018). *Report on The Set of Main Indicators for Assessing the Level of Business Development Throughout the Country and Localities in 2017 and the Period of 2010–2017*. Hanoi: General Statistics Office.

Hoang Van Tuyen. (2017, April 4). Technology and Innovation System in Developing Countries: Vietnam Case. *Vietnam Science and Technology Journal*, pp. 4–9.

Institute of Policy and Management. (2019). *Report on Management Policies on Social Mobility of High-Quality Science and Technology Human Resources of Vietnam in the Context of International Integration*. Research Results of the National-Level Project KX01.01/16-20, Hanoi.

Kraemer-Mbula, E. and Wamae, W. (2010). The Relevance of Innovation Systems to Developing Countries. In: E. Kraemer-Mbula and W. Wamae, ed., *Innovation and the Development Agenda*. Paris: OECD.

Mai Ha, Hoang Van Tuyen, Dao Thanh Truong. (2015). *Enterprise of Science and Technology*. Hanoi: Science and Technics Publishing House.

National Assembly of Vietnam. (2012). Higher Education Law.

National Assembly of Vietnam. (2013). Law on Science and Technology.

National Office of Intellectual Property of Vietnam. (2017). *Annual Intellectual Property Report 2017*. Hanoi: National Office of Intellectual Property of Vietnam.

Nguyen Van Anh, Khuat Duy Vinh Long, Le Vu Toan. (2012). *About Activities of the Local Science and Technology Development Fund*. Available at: https://thongtinphapluatdansu.edu.vn/2012/05/10/10-05-2012/.

OECD. (2014). *OECD Reviews of Innovation Policy: Science, Technology and Innovation in Vietnam*. Paris: OECD.

Thai Nguyen University. (2018). *2017 Annual Report of Thai Nguyen University*. Available at: http://tnu.edu.vn/thong-ke-cb-gv-dhtn-dn182.html.

Vietnam Academy of Science and Technology. (2017). *Operation Report in 2017 of the Vietnam Academy of Science and Technology*. Hanoi: Vietnam Academy of Science and Technology.

Vietnam Ministry of Education and Training. (2017). *Report on Science and Technology Activities in Higher Education Institutions in the Period of 2011–2016 and Proposed Solutions for Development in the Period of 2017–2025*. Hanoi.

Vietnam Ministry of Science and Technology. (2014). *Vietnam Science and Technology White Paper 2013*. Hanoi: Science and Technology Publishing House

Vietnam Ministry of Science and Technology. (2015). *Circular No. 03.2015/TT-BKHCN Issued on 09 March 2015, Regulating the Sample Charter on Organization and Operation of Science and Technology Development Funds of Ministries, Provinces and Centrally Run Cities*. Hanoi.

Vietnam Ministry of Science and Technology. (2016). *Vietnam Science and Technology White Paper 2015*. Hanoi: Science and Technology Publishing House.

Vietnam National University Ho Chi Minh City. (2014). *Report on S&T Research Activities*. Ho Chi Minh City.

http://ncstp.gov.vn/vi/gioi-thieu/chuc-nang-nhiem-vu.

http://scientometrics4vn.com/category/recent-posts/vietnam-vs-asean/2018-vietnam-vs-asean/.

http://www.pvn.vn/.

http://www.viettel.com.vn/.

CHAPTER 4

An Analysis of Vietnam's STI Policies in the Context of International S&T Integration

Vietnam is not an oasis in a globalized world—its STI policies also need to be integrated globally. In terms of system theory, the condition for policy integration is an isomorphism and not necessarily a homomorphism of the paradigm of policies. According to Vu Cao Dam, the paradigm has a four-level structure: (1) philosophy, (2) viewpoint systems, (3) normative systems, and (4) conceptual systems. Firstly, policies must have the same philosophy. At the moment, the world economy is an open one in which no country can close its economy, especially after Vietnam joined the WTO. In a market economy, enterprises must compete and constantly innovate themselves. The philosophy leads to diverse viewpoints due to specific conditions. The next step is the development of normative systems with local cultural differences and traditions. Finally, conceptual systems are formed, depending on the local tradition. Because of differences in the perspectives, standards, and conceptual systems, there is a need to create isomorphic paradigm policies.

4.1   STI Policies in Contemporary Vietnam

4.1.1   Tax Incentive Policies

Tax incentives are special forms of taxation policy for organizations, individuals, or businesses (taxpayers) when they meet certain conditions. In business, tax incentives create a comparison of tax-related benefits among

© The Author(s) 2019

D. T. Truong, *Perspectives on Vietnam's Science, Technology, and Innovation Policies*,
https://doi.org/10.1007/978-981-15-0571-3_4

taxpayers. The overall objectives of this policy are to attract investments, promote regional development, job creation, and SMEs' development.

Tax incentives have the following instruments:

- *Low standard tax rate*: The law specifies standard tax rates. If low standard tax rates are applied, it will attract more capital.
- *Preferential tax rates*: These are lower than the standard tax rates, prescribed by law, and are usually applied to specific beneficiaries.
- *Exemption of all or part of corporate income tax*: Decree 218/2013/ND-CP stipulates that enterprises are entitled to corporate income tax incentives, among which the preferential tax rate is 10% in a 15-year term for (1) newly established enterprises in remote areas, and (2) enterprises in specific fields of S&T, environment protection, and so on.
- Rapid depreciation is a form of tax incentive that helps enterprises to accumulate more capital because the deadline for tax payment can be delayed.

In Vietnam, SMEs account for 97% of the total number of enterprises, and they play an important role in creating jobs and increasing income for workers, as well as help mobilize resources for development and poverty reduction. Tax reduction and tax extension policies have been implemented, but these only address immediate difficulties for the businesses. In order to promote the effectiveness of tax policies for long-term development objectives, the state should focus on certain measures that will be explored below.

Currently, the tax rate that is applicable to all businesses is 25%, irrespective of the scale of enterprises. The preferential tax rates of 20% and 10% are applied based on specific industries and sectors. In order to promote the development of SMEs, the state should offer a lower preferential tax rate. Reducing the tax burden through tax incentives will increase profits for business and encourage more capital flows into the country.

The tax incentive policies are mainly import tax exemption and value-added tax incentives in the import of machinery at a tax rate of 0%. In order to encourage enterprises to invest in technological equipment renovation, the state should continue reducing import tax exemption and reduction for machinery. Article 16 of Decree 124/2008/Gov-Decree specifies tax reduction for different types of enterprises, such as a four-year tax exemption and 50% tax reduction for nine subsequent years for newly established enterprises in remote areas. Within a four-year period, enterprises are tax-exempted from the profits earned by the application of the research results. Hence, enterprises can reinvest the earning into production and renovate technologies.

Regarding value-added tax, the Law on Value Added Tax (VAT) No. 13/2008/QH12 has made some adjustments to create more favorable conditions for enterprises. Article 8 specifies the tax rates of 0%, 5%, and 10% for certain types of products.

The extension of tax payment time for SMEs is a necessary solution, filling the need for capital. In addition, it is necessary to reduce administrative procedures to help enterprises. Reasonable tax incentives bring many benefits to businesses. The reduction of taxes will help enterprises improve their competitiveness and strengthen their financial resources so as to invest more in technological innovation in production.

4.1.2 Policy to Promote Technology Innovation Capacity of Enterprises

Along with socioeconomic development, increasing human needs require innovation in qualities and services. As a result, technological innovation is considered an indispensable part for enterprises to survive and compete in the market. If companies do not innovate, their existing technology systems, machineries, and equipment will be outdated and ineffective, resulting in a limited quality and quantity of output. Technological innovation will help businesses improve product quality, expand the market for the product, and improve the position of the business.

4.1.3 Policy to Promote the Linkage Between Business and Researcher

4.1.3.1 Formation of Spin-Offs in R&D Organizations

There are many factors leading to the formation of the link between research and production, science and business, including:

- Scientific research is not just a fundamental research project but a business product. Therefore, enterprises must translate scientific ideas into production.

- By recognizing the importance of S&T in production, many enterprises are more active in research activities.
- Scientific research becomes an industry, so research institutes must change their organizational structure and operate as enterprises.

Spin-off enterprises are set up to translate scientific research into production and business. These enterprises are more effective with applied technologies. At present, some university research institutes have turned to spin-offs, which have helped commercialize technologies into production and business. In order for these organizations to be more effective, enterprises need preferential policies. Funding support can be extracted from venture capital funds. During operations, businesses will inevitably face difficulties. This support will help S&T enterprises to operate more stably.

In business activities, enterprises spend money to research and set up S&T organizations to carry out research activities, and this should be encouraged. The Vietnam Science and Technology Development Strategy 2011–2020 sets the following target: "By 2020, 60 basic and applied research organizations will be established to reach the regional and world level, capable of solving [the S&T problems of national importance]: 5,000 science and technology enterprises; 60 hi-tech incubators, hi-tech enterprise incubators." But with the current results of scientific research, it is difficult to achieve this goal because very few research topics are translated into production and business.

The priority policies for enterprises to implement or to collaborate with scientific organizations at the provincial and national levels should be practical, so that they can contribute to socioeconomic development and local security. At the same time, financial support for the implementation of research projects and the application of new technological advances for production and business of enterprises should be advanced.

4.1.4   Market Development Policy

Nowadays, market research is necessary. It requires the participation of governmental departments, specialists, and market analysts. In addition, business is an important factor in developing the market. The transfer of research results needs to expand the link between R&D human resources and businesses to facilitate mutual benefits and avoid competitive pressure from other stakeholders.

Technology is part of research output, while technology markets play an important role in socioeconomic development and technology transfer.

4.2 Vietnam's STI Policy Through the Development Stages

The current STI system in Vietnam is a continuation of the S&T management system of the Socialist Republic of Vietnam beginning in 1959—from the establishment of the State Committee for Science; to the Ministry of Science, Technology and Environment; and to the current Ministry of Science and Technology (Dao Thanh Truong 2016a) (Table 4.1).

4.2.1 From 1959 to 1981

The year 1959 witnessed the establishment of the State Committee for Science. This marked the formation of the S&T system in North Vietnam, a central economic planning system based on the socialist model of the Soviet Union with the following characteristics:

- The business function is solely manufacturing; there is no technology research department in enterprises.
- Research institutes are placed under ministries. These institutes work with the research plan designed by ministries and have no contact with businesses.
- Established universities have the sole function of training because the research is done by research institutes under ministries.
- Fundamental research is mainly carried out by the State Committee for Science, the Institute of Social Sciences, followed by the Institute of Natural Sciences and Technology, separated from the State Committee for Science. Recently, these institutes have been converted into academies. In addition to the basic research function as defined in the institutional and operational decisions of the institutes, in practice, these institutes also have a training function.

The model of Vietnam's STI system was formed before 1981. The characteristic of this model is that scientific research is separated from production and regulatory policies. This is reflected in the fact that all activities of production and scientific research are solely in accordance with the state's plan.

Table 4.1 System of policies on STI in Vietnam through periods

Period code	Related policies	Policies for S&T	Features of STI system
1.	"Fence breaking/Pha rao" period (1979–1986)	• "Pha rao" strategy 7/1979 • Instruction 100-CT/TW in 1981: extending production quotas to groups of laborers and individual laborers in agricultural collectives • Decision 25/CP, 21 January 1981: instructions and measures for improving the activeness in business activities and financial autonomy in state enterprises • Renovation Policy in December 1986	• Decision 175 in 1981: allowed R&D contracts (providing a platform for other renovation policies) • Decree on patents in 1981	• The state applied these policies to state institutes; universities and enterprises were excluded. • Strict system of management was top-down, which allocated resources and analyzed R&D and educational activities.
2	"Economic liberalization and reform" period (1987–1995)	• Foreign investment law, 29 December 1987 • Law for bankruptcy in 1988 • Land law in 1988 • Establishment of banks on two levels (1990): state banks and sub-commercial banks • Private business law in 1990 • Privatization process (1992) • State-owned enterprise law (1995) • Vietnam's participation in the ASEAN Free Trade Area (1995)	• Decision on monopolous management for S&T (1987) • Legal order on foreign technology transfer (1988) • Legal order on intellectual property law (1989) • Decree 35: allowed the establishment of R&D private enterprises • Decision 268: allowed the commercialization of R&D activities • The civil application of the intellectual property law (1995)	• State management primarily in state institutes, loose relations among the state, universities, research institutes, and businesses. • R&D organizations were allowed to establish relationships and contracts with individuals and non-state organizations • The National Center for Natural Sciences and Technology was founded and was in charge of conducting basic and applied research. • Budgets for S&T were provided by the state. • The state was the center in the STI system. • Universities played a blurring role.

(continued)

Table 4.1 (continued)

Period code	Related policies	Policies for S&T	Features of STI system
3	Restructuring of S&T system and development of S&T policy framework (1996–2000)	• Decree 57: extended the rights for export to domestic enterprises (1998) • Business law (1999) • Amendments of foreign investment law (2000) • The operation of Ho Chi Minh City department of stock exchange • Vietnam–US commerce agreement (2001)	• Decision 782: restructuring public R&D organizations (1996) • The establishment of Hoa Lac High Tech Zone (1998) • Decree 119: foundation of business support fund and land lease for S&T activities • Science and technology law (2000) • Decision 850: foundation of 19 central labs (2000) • Sai Gon High-Tech Zone (2002)	• The state as the central entity in the management of the STI system. • Universities played a blurring role. • Gradual development of the relation between R&D and production. • Development of infrastructure for development (e.g. high-tech zones).
4	Global integration, reform in state management of S&T, capacity building for policymaking (2003–2010)	• Joint investment with state budgets (2005) • Competition law (2005) • Education law (2005) • Amendments of the business law (2005) • Ha Noi stock exchange department (2005) • Participation in the WTO (2006) • Investment law (2006) • Five-year plan for SMEs development (2006) • Personal income tax law (2008) • Telecommunications law (2009) • Decree 56: support for SMEs (2009) • Project 30 for bureaucracy eradication (2010)	• Strategy for S&T development to 2010 (2003) • NAFOSTED (2003) • The national council of science and technology (2003) • Decision 171: reform on state management of S&T (2004) • Decision 214: technology markets (2005) • Decree 115: autonomy and self-accountability in public S&T organizations (2005) • Intellectual property law (2005) • Law of Tech Transfer (2006) • National foundation for technology transfer (2006) • Standard and technological norm law (2006) • Amendments to the intellectual property law (2009) • High-tech zone law (2008) • Division of technological applications and renovations (2007)	• The state was the central entity in the management of the STI system. • The components in the system interacted with each other, albeit their interactions seemed to be individual-focused. • There were policies on finance and facilities.

(continued)

Table 4.1 (continued)

Period code	Related policies	Policies for S&T	Features of STI system
5	Toward a fully fledged and functional innovation system (2011–2013)	• The National Assembly adopted the 2011–2015 socioeconomic development plan • Decree 20-NQ/TW, 31 October 2012: S&T development for industrialization and modernization in accordance with the market-socialist-oriented economy and international integration • Decision 418/QD-TTg, 11 April 2012, by the prime minister: approval of the strategy for S&T development in the period 2011–2020 • State's action program of S&T (Decree 46/NQ-CP) (2013) • Decision 2017/QD-TTg, 8 November 2013, by the prime minster: approval of the strategy for the development of the S&T market to 2020	• Decree 80: new regulation for cooperation in S&T investment (2010) • Decision 4009/QD-BKHCN: approval of human resource planning from 2011 to 2020 by the minister for S&T, issued on 29 December 211 • Strategy for S&T development from 2011 to 2020 (April 2012) • Division for S&T market and enterprise development (2011) • S&T law (2013) • Decision 677, 10 May 2011, by the prime minister: approval of the national reform in technology to 2020 • Decision 735/QD-TTg: approval of the national project "International Integration in Science and Technology to 2020" (2011) • Decree 20/NQ/TW: S&T development (2012) • Decision 592/QD-TTg: approval of the support programs for business development and S&T organizations—the practice of the autonomous and self-accountable mechanism (2012)	• Enterprises were the center of the STI system. • There were interactive relationships between the components.

(continued)

Period code	Related policies	Policies for S&T	Features of STI system
6	Toward an officially reformed system with perfect functions (2013–2015)	• Decree 99/2014/ND-CP: investment in potential development areas and encouragement for S&T activities at universities • Joint circular 55/2015/TTLT-BTC-BKHCN by the Ministries of Finance and S&T: Guidelines for building and distributing estimation and settlement funding for scientific and technological missions that use state budget	• Decision 2017/QD/TTg, 8 November 2013, by the prime minister: approval of the S&T market development program to 2020 • Decree 40/2014/ND-CP: regulations for the mobilization and recognition of human resources in S&T activities • Decree 87/2014/ND-CP: regulations for skill attraction in S&T activities to Vietnamese expatriates and foreign experts participating in S&T activities in Vietnam • Decree 99/2014/ND-CP, 25 October 2014, with effect since 15 December 2015: potential development areas and encouragement of S&T activities in universities • Decision 2395/QD-TTg, 25 December 2015, by the prime minister: approval of the project for human-capacity building in Vietnam or overseas through state budgets • 2014 business law • Decision 2245/QD-TTg: approval of the project for restructuring S&T branches until 2020 and vision until 2030, focusing on transforming development models to contribute to economic development (2015) • Decision 1747/QD-TTg by the prime minister: approval of the support program for applications of S&T advances to speed up socioeconomic development in mountainous areas and areas with ethnic minorities from 2016 to 2025 (2015) • Decree 16/2015/ND-CP: autonomy in public civil organizations	• The components in the STI were linked. • The relations between training, research, and research production were formed.

(continued)

Table 4.1 (continued)

Period code	Related policies	Policies for S&T	Features of STI system
7	The S&T system with the spirit of creative start-ups (2015–2018)	• Decision 142/2016/QH13, 12 April 2016: five-year plan for socioeconomic development from 2016 to 2020 • Decision 2245/QD-TTg: project of restructuring S&T branches till 2020 and vision till 2030 • Directions for S&T development in accordance with the approved decisions at the 12th National Party Congress (2016) • Decree 54/2016/ND-CP: autonomous mechanisms of public S&T organizations • Decree 35/NQ-CP, 16 May 2015: support and development program for enterprises till 2020 • Decision 844/QD-TTg by the prime minister: support for the national ecological system start-up and innovation projects till 2025 (2016)	• Decree 54/2016/ND-CP, 14 June 2016: autonomous mechanisms in S&T organizations • Decision 1381/QD-TTg by the prime minister, 12 July 2016: amendments and additions to Decision 592/QD-TTg on the support programs for S&T enterprises to develop and perfect technologies producing new products with competitive advantage in market • Circular 08/2016/TT-BKHCN: amendments and additions to Circular 32/214/TT-BKHCN on the S&T development program till 2020—additions to activities of the commercialization of research results and development of technology and intellectual property • Law on support for SMEs in 2017 • Technology transfer law in 2017 • Decision 844/QD-TTg by the Prime Minister on the support for the national ecological system start-up and innovation projects to 2025	• Start-up movements and spirit for creativity were being formed with proper care. • The components in the STI system were naturally attached to each other in response to market demands.

Notes: Periods 1–5: According to OECD (2014, p. 23); Periods 6–7: According to Dao Thanh Truong (2016b) and Institute of Policy and Management (2019)

Research funds of the institutes come from the state budget and are granted under the principle of "earning sufficiently and spending sufficiently". If the expenditure is insufficient, additional estimates will be made for the budget allocation. If there is excess spending, the budget is returned and all processes are remitted to the budget.

On the business side, profits are "distributed" by the state according to the "norm profit", which is calculated as a percentage of the cost. For example, the percentage for light industry is 5% and heavy industry is 8%.

At this stage, Vietnam's STI system has some the following characteristics:

- Research and production are not shared-interest. The research results from the institutes are submitted to the state, which then reapplies the research results to another unit. The institute has no responsibilities and interests.
- The higher the production cost, the greater the profit distribution, so the producers are not interested in technological innovation.

On 1 December 1978, the Government Council issued Decree No. 302-CP promulgating the Regulation on the Union of State-owned Enterprises, and the prime minister issued Directive No. 548-TTg of 1 December. The latter instructed the organization of production unions into national economic industries, which aimed to overcome the division of production. In addition, between the 1980s and 1990s, there was the emergence of scientific production unions in order to create the linkage of the R-D-P-M elements. This is an indispensable development of the STI system in a command economy with the desire to innovate products and have technological innovation to develop the economy.

4.2.2 Decision No. 175/CP, 1981

The transformation of the Vietnamese economy began in 1981 with Directive 100 of the Central Secretariat of the Communist Party of Vietnam on the regime of final product allocation to households and workers in agriculture. This was followed by Decision 25/CP and 26/CP on the three-part plan in industry. As a result, the STI system also entered the reform process with Decision 175/CP of 1981, allowing R&D organizations to collaborate with each other and with businesses. Vietnam's STI system has undergone major reform since the 1970s, as it takes on the characteristics of a market economy. The nature of the reform was that, in addition to the task of carrying out research plans assigned by the state, independent research institutions and research institutions located in universities were allowed to sign research agreements and apply the research results into production. A big turning point in the STI system was when the government allowed S&T organizations to sign contracts according to Decree 51/CP on the diversification of scientific and technological activities. From that, they are able to implement their own research plan based on market demands.

Following Decision 175/CP, there were a number of accompanying regulations, notably regulations on the source of funds obtained under the contract:

1. *Circular 1438/TC-KHKT:* the right to use funds collected under the contract, not to be remitted to the budget as before.
2. *Circular No. 1438/TC-KHKT:* funds collected under the contract are classified into three types: (i) scientific development fund; (ii) reward fund; and (iii) welfare fund. R&D organizations actively use these funds.
3. *State Bank of Vietnam's Circular No. 451/KHKT-NH:* allows S&T organizations to open accounts to spend income amounts on contracts. This was a step forward because scientific organizations had not been allowed to open their own types of equity accounts.
4. *State Bank of Vietnam also issued Directive 16/NH-CT:* lending to S&T organizations to implement the contract.

Financial and banking regulations for S&T organizations, such as setting up a fund and opening a deposit account, are considered a revolution as universities and institutions were considered as "administrative agencies" in the non-material production area.

Thus, Decision 175/Gov in 1981 created a major philosophic turning point in Vietnam's STI system. It led to the change from a centrally orientated STI to a decentric orientation with a horizontal contractual relationship in the link between the RDPM elements.

4.2.3   Resolution No. 51/Minister Council, 1983

If Decision 175/Gov created a horizontal relationship in the STI system, Resolution 51/Minister Council issued on 17 May 1983 was a new milestone as it facilitated the formation of spin-offs in the STI system. A number of centers were set up in institutes and universities to serve production from research results.

Resolution 51/Minister Council provided an opportunity for R&D organizations to proactively target the market, creating demand for the market. A number of centers were established at the Vietnam Academy of Science, Hanoi University of Technology, and Ho Chi Minh City Technology University. The nature of these centers is spin-off, spin-in, spin-out, start-up, and so on. These types of businesses have the function of supporting research ideas, and their technology and products make up the market. S&T organizations—such as Kova Paint Company, FPT Corporation, and Thermal Engineering Center—are spin-off enterprises that have been transformed into start-ups and have become efficient enterprises. This is a very important step to the formation and promotion of the commercialization philosophy of R&D in the market.

4.2.4 Decision No. 134/Minister Council, 1987

Decision No. 134/Minister Council was issued on 31 August 1987 with two noticeable points:

Firstly, the state allowed market price agreements in technology transfer, creating incentives for technology to be replicated in society, avoiding the monopoly of production.

Secondly, Decision 134/Minister Council allowed researchers to enter into contracts.

4.2.5 Ordinance on Technology Transfer, 1988

The Ordinance on the Transfer of Technology from Abroad into Vietnam, or the Ordinance on Technology Transfer, in addition to the provisions governing the relations in technology transfer, is another regulation that is considered a philosophic turning point. With the law, the private sector now has the right to take part in technology transfer contracts. While the Foreign Investment Law published in 1987 did not allow private participation in technology transfer, the amendment of the law in 1990 opened the activity up to the private sector. The Ordinance on the Transfer of Technology in 1988 created a new philosophy in Vietnam's STI system for international integration and was the beginning of the privatization of STI activities.

4.2.6   Decree 35/Minister's Council, 1992

Decree 35/Minister's Council was issued on 28 January 1992 with some important features:

First, all citizens have the right to conduct S&T activities and establish S&T organizations.

Second, the establishment of S&T organizations only needs to be registered in one competent organization, without complicated permission, and the conditions for the establishment of the S&T organization are the written decision of the founding entity.

Third, private S&T organizations have the same rights as state-owned S&T organizations in conducting S&T activities, including participation in bidding for research tasks set by state institutions.

Finally, human resources in research areas are under long-term contractual regimes and have the same rights as those employed by the state.

Decree No. 35/Minister's Council of 1992 was a great step forward in democratic rights in S&T activities. It ensured that all people have the right to organize and operate S&T. The philosophy of this decree is the democratization of S&T.

4.2.7   Decree 115/2005/Gov-Decree

Decree 115/2005/Gov-Decree (abbreviated as Decree 115), issued on 5 September 2005, is another milestone in the reform of S&T policy philosophy as it moved the system from the science of the state to autonomous science. If this philosophy was implemented, then the STI system of Vietnam would be fully compatible with the context of integration. However, over the 10 years of implementation, Decree 115 has not been realized. According to Vu Cao Dam, this is caused by the lack of consistency between purpose and concrete means. The decree aimed to allow public S&T organization to be self-reliant and self-responsible, but not autonomous in proposing research tasks. Moreover, these organizations are autonomous in seeking out-of-the-field contracts, but in reality, there is no market economy. In the university sector, better results were achieved with the implementation of autonomy and self-responsibility under Decree 43/2006/Gov-Decree on 25 April 2006 (Decree 43). This was replaced by Decree 16/2015/Gov-Decree on 14 February 2015.

4.2.8 Law on Science and Technology (2013)

The Law on Science and Technology 2013 published on 18 June 2013 recorded and amended some terms assigned in the provisions of the Law on Science and Technology issued in 2005. The Law on Science and Technology 2013 expressed the innovative view regarding the mechanism for the management of S&T activities, in line with Resolution No. 20-Decree/Central of the Sixth Plenum of the 11th Party Central Committee on Science and Technology Development, in service of the industrialization and modernization of business conditions in a market-oriented socialist economy and in the context of international integration.

Highlights of the Law on Science and Technology 2013 include the financing of S&T activities such as training talents. Article 49 of the Law stipulates that the state shall ensure that expenditure for S&T of 2% or more of the total annual state budget expenditures will be gradually increased according to the development requirements of S&T. It is noteworthy that this law also mentions the purchase of R&D results, support for high-tech imports, and application of the mechanism of package expenditures for S&T activities using the state budget.

Regarding S&T human resources, the law set specific regulations on the training, retraining, and attracting of S&T human resources, with provisions relating to treatment regimes in three aspects: research environment, income/living conditions, and honors.

In addition, the law is considered to be "updated" as it introduced some regulations to promote international integration in S&T, such as building S&T research groups with international standards; strengthening the national database system on S&T; establishing key laboratories with regional and international standards; perfecting preferential mechanisms, policies, and support for Vietnamese organizations and individuals to participate in international integration activities in S&T; and formulating mechanisms and policies to attract foreign organizations and individuals to participate in Vietnam's S&T development.

4.3   VIETNAM'S STI POLICY IN THE TREND
OF INTERNATIONAL INTEGRATION

International integration is an ongoing process with philosophical changes, beginning with Decision 175/Gov in 1981. The state-led system has transformed into a multistakeholder system with common standards of the international S&T community. The nature of the transition is in line with the transformation from a state-led economy to a multisectoral market economy—a "socialist-oriented market, regulated by the state", as mentioned in the Communist Party of Vietnam's Sixth Congress document. Therefore, the STI system fully meets the conditions necessary for the integration process, including the following:

First of all, Vietnam is moving from a state-led economy to a socialist-oriented market economy with many economic sectors involved. The STI system is no exception—it also becomes a multicomponent STI system, as announced in the 1988 Ordinance on Technology Transfer and Decree No. 35/1992/Minister's Council.

Second, in the world today, state management is inevitable. However, the concept of state management is understood very differently in different countries. There is a view that state management means that the state will manage and participate in all activities. There is also the view that the state governs only at the macro level. In order to integrate, it is necessary to set up macroscopic institutions for state management of S&T to be compatible with other countries worldwide

4.3.1   *Analysis of Vietnam's STI Policy*

Generally, Vietnam's policies can be said to comprise three levels, with each level subdivided into modules.

1. *Level 1: general objective of the policy system.* The philosophy of the whole economic system. It is a socialist-oriented market economy with the participation of many economic sectors whose rights are equal in this open system to the world market.
2. *Level 2: including Level 2 systems.* Viewpoints of policy instruments such as financial, labor, and wage policies. All these policies follow a market-oriented philosophy.

3. *Level 3: including Level 3A.* Viewpoint system about S&T activities and innovation activities. The entities that organize these activities are research institutes and universities. These modules ensure that they follow the market-oriented and multistakeholder philosophy.
4. *Level 3: including Level 3B.* Viewpoint system of the regional area, including manufacturing and business enterprises.
5. *Level 3: including Level 3C.* Viewpoint system of auxiliary activities, which ensure the infrastructure of innovation, including information, technology, and industrial infrastructures.

4.3.1.1 Level 1 Policy Status

In Vietnam, a study has shown that the share of investment in state-owned enterprises accounted for more than 40% of its GDP, but state-owned enterprises only account for 32% of the GDP. Meanwhile, investment in non-state sectors account for only 38% of the GDP, but the proportion of added value by the sector was up to 49%. Particularly, the share of GDP in the private sector constitutes 33%.

A comparison of the incremental capital-output ratio (ICOR) coefficients presented in Table 4.2 shows that the non-state sector is more efficient than the state sector.

4.3.1.2 Level 2 Policy Status

When it comes to financial policies, it is clear that STI activities in Vietnam are still far from being integrated internationally.

- First of all, Vietnam does not have many funds for innovation, especially venture capital funds. There are several venture capital funds in Vietnam such as Viet Capital, but the basic principle for the operation of venture capital funds in Vietnam is almost nonexistent. The

Table 4.2 ICOR figure measured by capital

Ordinal number	Stage	2000–2010	2000–2005	2006–2010
	Total	6.07	4.89	7.45
1	State economy	8.53	6.94	10.85
2	Non-state economy	3.28	2.93	3.55
3	Economy with investment	9.65	5.20	15.26

Source: General Statistics Office

2013 Science and Technology Law has a clause on venture capital funds, but the funding is based on state budget investment, which is not really suited for the characteristics of venture capital funds. This is similar to the formation of the National Hi-Tech Venture Capital Fund defined in Section 2, Article 25 of the High Technology Law 2006: *The financial source for the formation of the National Hi-Tech Venture Fund consists of the charter capital of the National Hi-Tech Venture Capital Fund formed from the state budget and replenished from the state budget in the course of operation.*

- On tax policy, there is no policy instrument for innovation or for the entire operation of the STI system.
- Human resource policies for STI activities in Vietnam have not really stimulated human resources to participate in S&T activities to create S&T achievements.

In summary, these policies do not focus on encouraging participation in S&T activities and creating S&T products.

4.3.1.3   Level 3 Policy Status

When analyzing the stakeholder groups that implement STI policies—specifically universities, businesses, research institutes, and subordinate units of innovation—these organizations are sources of achievements for Vietnam's STI system in the trend of international S&T integration. Essentially, these components have to be directly involved in integration and must have an isomorphic paradigm to the world's STI community. However, in general, these components are not isomorphic in paradigm.

(a) *Universities*

Universities in Vietnam only have the function of training, and scientific research activities are considered part-time and are done through projects and programs. Recently, a number of universities established research institutes in universities such as the Hanoi University of Science and Technology. However, research activities do not see sufficient investment.

At present, Vietnam is researching and implementing the construction of the "research university". However, the general regulations for research universities are not available. In Thailand, for example, this issue has been implemented as the National Institute for Development Administration

(NIDA). Although it is an institute, it has the same function as a university. It can be seen that Vietnamese universities nowadays still operate in a paradigm that is quite different from that of other universities in the world and does not meet the conditions for integration.

(b) *Research institutes*

It is possible to divide research institutes in Vietnam into four categories:

The first is governmental institutes, which are independent academies modeled after those in the former Soviet Union. In the past, these institutes were merely research activities without training or production activities. A model of an academy has recently emerged that links academic training and research.

The second type is institutes belonging to ministries, which are also independent academies according to the former Soviet Union models. These institutes are not linked to training and production; they are institutes created for business purposes by the production and business units.

The third type is institutes located in universities. These have recently appeared since Vietnam began making changes. These institutes belong to the paradigm of the contemporary world's STI system, which suits the needs of integration, but they are still weak.

The fourth type is the sector of institutes in the manufacturing corporation. This type has emerged since Doi Moi. These institutes also belong to the paradigm of the contemporary world's STI system, but they are still weak.

From the above analysis, it is possible to comment on the network of research institutes in the integration trend. Vietnamese institutes, despite many changes, still do not meet the conditions for integration.

(c) *Businesses*

Business is the basic unit to create innovative achievements in the form of new products and technologies. In all industrialized countries, businesses have an R&D unit.

In Vietnam, R&D activities are carried out (separated from production) by an R&D institute located within the ministry and independent of the enterprise. Since the development of the market economy, many manufacturing corporations with R&D units have emerged in a number of industries. However, according to the author's survey, there are quite a lot of research institutes that are not really associated with the operation of the corporation. Only some institutes such as the Petroleum Institute under the Petroleum Group operate efficiently.

Enterprise Law No. 60/2005/Congress 11 issued on 29 November 2005 created a very high degree of autonomy for businesses, including the right to self-organize R&D units. On 26 November 2014, the National Assembly passed the Enterprise Law of 2014. This law has many additional points that are more advanced than the Enterprise Law 2005. The author conducted a number of interviews with business owners, who stated that businesses are claiming autonomy in reforming and forming innovative organizations. However, enterprises cannot be reformed because Vietnam does not have a complete market economy. For example, the tariff barrier protecting the domestic automobile industry has made Vietnam's automobile technology come to a standstill, with no impetus for innovation.

At this point, it can be seen that the innovation mechanism in the enterprise is entangled in a paradox in which enterprises are opened up with a great initiative for innovation, but the environment is not ready to innovate into a perfect market. This paradox is similar to the paradox caused by the effects of Decree 115.

4.3.2   Relevance of Vietnam's STI System to the Trend of International S&T Integration

From the above analysis, it is possible to conclude as follows:

Firstly, Vietnam's STI system is *quite distant* compared to the world's STI system. This system is structured with the following specific characteristics:

- *Scientific research is separate from training.* Vietnamese universities have not promoted research functions, and academic institutions have not linked research with training. A number of universities now have research institutes, but state regulations have not kept up.
- *Manufacturing is separated from research.* Most businesses purchase and use old technologies, and they are not interested in technological innovation.

Second, Vietnam's policies have begun to focus on innovation, but they are fragmented: (1) Decree 115/2005 /Decree-Gov for institutes; (2) "Academies" stand apart; (3) Decree 16/2015/ ND-CP on autonomy and self-responsibility in universities has just replaced Decree 43/2006/ND-CP; (4) Enterprise Law No. 68/2014/ Congress 11 dated 26/11/2014 for the manufacturing sector. There is no indication of integration of these four areas into an STI system.

Thirdly, Vietnam does not really have a perfect market economy. This is the basic factor to put pressure on innovative reform. The market expresses the supply and demand relationship of innovation whereby enterprises need to compete, innovate, and exploit innovative supply from the R&D sector.

Fourthly, S&T policies are still "pushing". According to this thinking, the whole STI system is always trying to create a market. However, this market does not create the need for technological innovation.

Fifth, the cause of underdevelopment *is from the S&T industry*—not only the supply side, but also from the demand and market sides.

Sixth, policy efforts aim to stimulate the supply side, but the demand is the decisive factor to fostering innovation. When there are supply factors but a lack of demand factors, the initiative and research from the supply side are external from the business. This is because Vietnam does not have a perfect market economy. The typical example is the automobile industry. In the automotive industry of Vietnam, there is no need for innovation. The condition for solving this problem is not inherent in the STI itself, but within the macroeconomic regulatory system.

After joining the WTO, the AEC and coming to join the TPP, it is imperative to enter the perfect market economy. The STI system must prepare policies for innovation and for the world's integration into the STI system. Vietnam needs to identify and formulate key arguments based on a different policy approach to readily embrace the integration process, in accordance with characteristics of the market economy.

REFERENCES

Dao Thanh Truong. (2016a). *Chính sách Khoa học, Công nghệ, và Đổi mới (STI) của Việt Nam trong xu thế hội nhập quốc tế: Thực trạng và giải pháp* [Science, Technology, and Innovation Policies of Vietnam in the Trend of International Integration: Situations and Solutions]. Hanoi: Thế Giới Publishers.

Dao Thanh Truong. (2016b). *Di động xã hội của nhân lực khoa học và công nghệ trong bối cảnh hội nhập quốc tế: Lý luận và thực tiễn* [Social Mobility of Science and Technology Human Resources in the Trend of International Integration: Theories and Practices]. Hanoi: Thế Giới Publishers.

Institute of Policy and Management. (2019). *Report on Management Policies on Social Mobility of High-Quality Science and Technology Human Resources of Vietnam in the Context of International Integration.* Research Results of the National-Level Project KX01.01/16-20, Hanoi.

OECD. (2014). *OECD Reviews of Innovation Policy: Science and Technology and Innovation in Vietnam.* Paris: OECD.

CHAPTER 5

Policy Recommendations to Promote Vietnam's STI System in the Era of International Integration

5.1 Restructuring Vietnam's STI System in a New Era of International Integration

Overall, Vietnam's STI system has the following characteristics:

First, the STI system is fragmented. There is a separation of the key components of the innovation process between training, research, and production, where most enterprises do not have R&D departments. S&T research institutes are separated from businesses and universities; institutes of basic science and institutes of social sciences and humanities are separated from universities; and universities purely focus on training and are separated from scientific research.

Second, each and every sector in the STI system seeks ways to "incorporate", but it is not efficient.

Third, the Vietnamese STI system is still dominated by the state. Although there has been a change in the level of management, the policies have not really worked.

The process of restructuring Vietnam's STI system is a two-stage process of reengineering:

Phase 1. Overcoming the barriers of the old system
Phase 2. Integration in the world's restructuring trend

© The Author(s) 2019   181
D. T. Truong, *Perspectives on Vietnam's Science, Technology, and Innovation Policies*,
https://doi.org/10.1007/978-981-15-0571-3_5

Essentially, Vietnam's economic reform is a process of transformation from a state-centric economy to a market economy. It is a process of economic restructuring in general, and STI system restructuring in particular. Besides transitioning to an STI system in a market economy, Vietnam's STI system must also collaborate with other countries to overcome the backwardness of the world's STIs—a double restructuring process.

The restructuring process of the STI system includes the following steps:

Step 1: Move from a state-centric STI system to a multistakeholder STI system and ensure equal rights for all components. This is becoming a reality, but the non-public sector is still very weak.

Step 2: Reinvent the relationship between research and production. At the same time, it must replicate the relationship between research and training.

Step 3: Integrate with the world's STI system.

The most common problem of reconstruction is the creation of an STI system with the participation of multiple stakeholders, to reconstruct the inherent relationship between science, training, and production. Based on the experience of building STI systems in many countries, the author proposes some options as follows:

Option 1: Raising salaries and research funding for universities in order to attract S&T personnel to the universities, thereby enhancing their research capability. Following this approach, Vietnam will have difficulties with the functional separation of the three components of the STI system: the university only has the teaching function, the institute only has scientific research functions, and the enterprise only focuses on production.

Option 2: Development of training activities in the academies. In fact, the Vietnam Academy of Social Sciences now trains S&T human resources. The Vietnam Academy of Science and Technology was also handed over by the Ministry of Education and Training to the Hanoi University of Science and Technology on 20 April 2016. This is a good opportunity to link training and research. This model should be promoted to link the two functions of training and scientific research to improve the quality of training and the effectiveness of scientific research, which will increase the applicability and transfer of scientific research results of universities from research institutes into practice. In this model, available resources on S&T human resources and the scientific research environment are utilized for training, scientific research, and production. At the same time, this method of integration will strengthen the capacity of S&T human resources, as university lecturers and researchers at research institutes would carry out both scientific research and training.

Vietnam is actively involved in the process of international integration. The restructuring process aims to create a compatible, qualified STI system. At present, the Vietnam's STI system does not satisfy the compatibility conditions in the integration. The compatibility of the STI model is reflected in the characteristics of a multistakeholder STI system, including state-owned universities/institutes as well as private institutions and foreign affiliated establishments. In addition, this integration is also expressed in S&T human resources with the active cooperation with the international STI community. In an autonomous STI system, the university is empowered to decide on training disciplines, enrollment standards, training programs, and international cooperation programs. Vietnam should soon sign up for the 1999 Bologna Declaration to form a higher education and science system that is closer to those of other countries, which will help Vietnam to integrate into the international STI community.

5.2   International Integration STI System in Vietnam

The model proposed in this section draws on the experience gained from other countries.

5.2.1   Form a Multistakeholder STI System

A multistakeholder STI system has been established since the adoption of Decree 35/1992, but in fact the private sector is still weak. The system does not have the potential to thrive. The restructuring process must be very diverse, including not only state-owned organizations and associations, but also private organizations. However, private STIs are not really equal to public STIs seeking funding for research, including state funds.

5.2.2   Recreate the Linkage Between Research, Training, and Production in the STI System

The link between universities, research institutes, and businesses has been a popular trend in the world and is considered an effective solution. The reality is that research plays an important role for universities to create new knowledge that contributes to S&T human resource training. It is important for businesses with technology transfer to apply technological achievements in production and to form a value chain within the STI system. With such importance, Resolution 2 of the 8th Central Committee Conference confirmed that:

> *"Universities must be centers for scientific and technological research, transfer and application of technologies to life"* and *"ensuring the combination of research institutes and universities, linking research—development with production—business"*. Conference 6 of the IX Central Committee set the task of *"improving the quality of education efficiency"*, *"improving the quality and efficiency of science and technology activities"*, and *"strengthening the association of universities with the Research institute and enterprise."*

In Article 1, Decision No. 418/QD-TTg of the Prime Minister dated 11 April 2012 on Science and Technology Development Strategy for the period 2011 to 2020, it stated:

> *To develop science and technology together with education and training is a top national policy, a key motive for rapid and sustainable national development. Science and technology must play the leading role in creating a breakthrough in production force, reforming the model of growth, enhancing the competitiveness of the economy, speeding up the industrialization process and modernizing the country.*

Universities, research institutes, and enterprises cooperate so as to use disseminate research results for teaching, as well as the service of production and business activities. The functions of these three stakeholders are complementary to each other. Research activities in research institutes will provide knowledge for teaching activities. Scientific results will be commercialized and marketed thanks to the production of the business. Training activities will contribute to the dissemination of S&T knowledge and S&T human resources for research institutes and enterprises.

Enterprises will apply research results directly to the production and business activities of enterprises.

In the context of Vietnam, research, training, and production functions are not merely the linkage between the symbiotic relationships between the institutional, school, and enterprise components in the STI system, but also the integration of these activities within the same organization. For example, the formation of spin-on, spin-out, spin-off, and startup technology enterprises in universities to implement scientific research results into practice, as the Hanoi University of Science and Technology and the Vietnam National University did. Likewise, the formation of universities in research institutes such as the Academy of Social Sciences from the Vietnam Academy of Social Sciences and the Hanoi University of Science and Technology under the Vietnam Academy of Science and Technology. Along with that is the formation of the university model in technology enterprises, such as the FPT University. This is a good premise, as they are the pioneering models that can create a strong link between research, training, and production in Vietnam's STI system. In addition, the reconstruction of this link will attract financial resources, training, and high-quality S&T human resources.

5.2.3   Forming a Truly Autonomous STI Institution

Being autonomous means not being controlled from the outside. In the 2000s, the state adopted a policy of autonomy and self-responsibility for enterprises and organizations called "non-business organizations". This was done with the issuance of Decree No. 115/2005/Decree-Government on the autonomy and self-responsibility of public S&T organizations; and Decree 43/2006/Decree-Government (now amended Decree 16/2015/Decree-Gov) on the autonomy and self-responsibility for public service delivery organizations, including the university. With the promulgation of these policies, in terms of macroeconomic management philosophy, the state recognizes the autonomy and self-responsibility of the S&T establishment.

Specifically, the incompatibility between self-reliance and self-responsibility for provisions for autonomy and self-responsibility should be dismantled. As a result, it is imperative to continue to adjust the philosophy of macro management so as to make Vietnam's STI system compatible with the world's STI system.

5.3   Policy Solutions for Promoting the International Integration of the STI System

In addition to reconstructing the STI system at the macro level, policy recommendations that promote STI activities within STI components are also critical in order to promote the international integration of Vietnam's STI system.

5.3.1   Policy Solution That Promotes STI Activities in Universities

Promoting STI activities in universities is one of the important ways to form a linkage between training and scientific research. This can be done by increasing investment in human and financial resources for research in the university sector, such as higher salaries and research funding, as well as formulating policies to create a legal corridor for the formation of research institutes in universities, thereby enhancing the research function for the universities (Dao Thanh Truong 2016). In addition, educational reform is a very urgent task, because education is a subset of the STI system. The world is reforming education according to the Bologna Declaration of 1999. For example, as early as the twentieth century, science in Vietnam was classified into Sector A, Sector B, Sector C, Sector D, and Sector H, while the world is now moving to a career-oriented system. Therefore, there can be no development in S&T without educational reform.

5.3.2   Policy Solution to Promote STI Activity in Research Institutes

Policies in research institutes can overcome the gap in training and production by the following measures:

- Invest in training function of the academies.
- Create research regulations for existing universities
- Modify the philosophy of Decree 115 to facilitate autonomy and self-responsibility of the institutes in accordance with the trend of scientific self-autonomy in developed countries.
- The state will invest in institutes of key technology development orientations, such as Korea's Korea Advanced Institute of Science and

Technology (KAIST) model, to enter priority areas for the country's future in the long term future.

5.3.3 Policy Solutions to Promoting STI Activity in Enterprises

Enterprises are the nucleus of innovation. Recreating the scientific and production linkages is a fundamental way for Vietnam's STI system to integrate into the world.

In the early 1980s, when the state created conditions for enterprises to make a three-part plan under Decisions 25/Gov and 26/Gov, and concurrently to institute signing contracts under Decision 175/Gov, technology institutes and technology universities responded actively. This was followed by Decision 51/Ministry Council allowing spin-off production and Decision 134/Ministry Council, which provides additional conditions for voluntary individuals and groups to promote further linkages between institutes/schools with manufacturing facilities. Institutions were thinking of self-accounting. The Ministry of Mechanics and Metallurgy even made decisions for enacting eight self-accounting pilot institutes. According to the Decree 115/2005/Decree-Gov, the institutes explained that they would become basic research institutes to take the subsidy regime in order to avoid self-reliance and self-responsibility. This has really become a problem. Therefore, the task of reconstructing the relationship between research and production is very urgent and difficult. There needs to be a mechanism for the institutes to self-transform, as Vu Cao Dam stated in his article in the *Journal of Science Activities* in 2000: "Where do the institutes of technology in our country go?" On such a basis, there should be policies such that the institutes would convert themselves into alternative options, in order for the organization itself to be the most profitable. Institutes can become businesses with many spin-off R&Ds. The premise for this model began after the government issued Decision 51/Minister Council in 1983 and Decision 134/Minister Council in 1987. On the other hand, institutes could become institutes of technology with spin-off production. This phenomenon has become popular with a number of institutes of technology in Vietnam that were recently established. Moreover, institutes can become engineering enterprises. In the present official documents, this type of technology enterprise is referred to as "S&T enterprise" and is defined as the type of enterprise that uses research results to produce goods. In addition, the institutes can become consultative organizations with a wide spectrum of activities ranging from legal consultancy, technology consultancy, and investment consultancy to construction contracting.

5.3.4   *Policy Solutions to Strengthen Links Between Components Within the STI System*

It is necessary to change the organizational philosophy of the STI system so that the market is the center of the S&T innovation system. The market will create competition, putting pressure on businesses to innovate. Businesses will then attract institutions/universities and the entire STI system to get involved in the innovation process.

Under a perfect market economy, the state will not manage economy, science, and education. The market will regulate the relationship between science and training at the macro level. The state will play the role of a donor, and funds will play a role in financing STI activities. The macro management functions of the state include: formulation of policies for the restructuring process; reconstruction of the links between science and education; replicating the relationship between science and production; implementing a policy exercising the autonomy of STIs; and replicating the policy structure of a macro management system. The state will publish and adjust policies, such as priority policies for opening new training disciplines and research directions as well as policies for STI autonomy and the inspection of policy.

REFERENCES

Dao Thanh Truong. (2016). *Chính sách Khoa học, Công nghệ, và Đổi mới (STI) của Việt Nam trong xu thế hội nhập quốc tế: Thực trạng và giải pháp* [Science, Technology, and Innovation Policies of Vietnam in the Trend of International Integration: Situations and Solutions]. Hanoi: Thế Giới Publishers.

Prime Minister. (2012). *Decision No. 418/QĐ-TTg of 11 April 2012 on Science and Technology Development Strategy for the Period 2011 to 2020.* Hanoi, Vietnam.

Vu Cao Dam. (2016). *Paradox and Escape.* Hanoi: The Gioi Publishing House.

CHAPTER 6

Conclusion

The S&T revolution in the world has been growing at an accelerated pace with breakthrough innovations that are unpredictable and have a huge impact in all aspects of human life. S&T is becoming the leading force in all fields. The strength of each country depends largely on S&T capacities. The advantage of natural resources and cheap labor are becoming less important. Meanwhile, the role of human resources with professional qualifications and creative abilities has become imperative in the context of economic globalization. The competitive advantage lies with companies that use new technologies to create new products and services that meet the varied and ever-changing needs of their customers. In order to adapt to the above context, developed countries are adjusting their economic structure toward fast-growing industries and services with high-tech and environmentally friendly technologies and services. Many developing countries prioritize training high-quality S&T human resources, increasing investment in research and technological innovation, especially in certain priority technology directions, and enhancing information infrastructure and communication to create competitive advantage and narrow the development gap.

Factors of the STI system include not only businesses, universities, and research institutes, but also other organizations such as associations and non-commercial organizations. The STI system consists of a set of components—M (Market)—R (Research)—D (Implementation)—P (Production)—that interact in pairs.

© The Author(s) 2019   189
D. T. Truong, *Perspectives on Vietnam's Science, Technology, and Innovation Policies,*
https://doi.org/10.1007/978-981-15-0571-3_6

Experimental studies from several countries including Northern Europe, Western Europe, Eastern Europe, Japan, South Korea, and some ASEAN countries show that some features of the STI system of other countries are:

- Increasing R&D expenditure, with a focus on priority areas of investment.
- Promoting spin-off businesses and startups to support SMEs, especially high-tech ones, by disseminating research results and support to them.
- Access to scientific research.
- Strong links between technology and science, with relatively high rates of public research funded by industry.
- S&T human resources are invested and trained to improve quality.
- Actively promote international exchanges and research programs.
- The role of government in building a comprehensive STI policy system that strongly supports human and financial resources, the development of various components of the S&T system, and national renewal.

The main lessons learnt are (1) linkage at different levels of the socio-economic development strategy with S&T development strategy in the framework of the national innovation system, which is enterprise-centric; (2) importance of creating an institutional environment conducive to the coherence and self-linking of S&T activities with production and business through the role of S&T enterprises as key actors in technological innovation; and (3) necessity to ensure the feasibility of strategic S&T development objectives through a combination of the role played by a number of strategic-oriented tools such as the development of a long-term vision and national technology roadmaps.

Through surveys conducted, the author notes the following characteristics of universities, research institutes, and enterprises—the three components of Vietnam's STI system. Universities play an important role in the creation of intellectual property, including research/invention, thereby contributing to innovation. Moreover, universities are the S&T human resource training center. Although many universities have achieved great success in technology transfer and commercialization of inventions, the overall contribution of S&T and technology transfer activities at Vietnamese universities is still low and not commensurate with the potential of a large team of scientists and researchers. A majority of the connections and interactions between the two entities—universities/research institutes and businesses—are spontaneous, scattered, and not systematic. Research institutes are heavily dependent on the state for their strategy, personnel, funding, and so forth, and only perform research functions. In general, the scientific workforce in enterprises is limited in number. The distribution of human resources is not equal, mostly dividing by region and fields of operation. The problem of skills shortage has not been overcome. Companies spend about 10% of their revenue on STI investments. The funding of S&T activities of enterprises is mainly from their own sources. It can be concluded that Vietnam's STI system bears the following three main points.

Firstly, there is the fragmentation and separation of the relationship between training, research, and production. Most businesses do not have R&D departments. S&T research institutes are separated from businesses and universities. The basic science institutes and the social science and humanities institutes are also separate from the universities, which purely teach.

Secondly, there is no real linkage among regions in Vietnam and between institutes and schools.

Thirdly, despite many changes, Vietnam's STI system is also bearing the characteristics of one that is in the command economy.

With the rapid development of S&T, the speed of technological innovation has become the basic determinant in the competitiveness of products, enterprises, and countries. Innovation is required to improve competitiveness in the market. Performing this task requires finances, human resources, production reorganization, and so on. The characteristics of STI activities in Vietnam show that there is very little involvement of the private sector, low investment, state inefficiency, and low output correlated with the needs of the economy and society.

Vietnam's STI policy is based on the state-run economy. There is no policy for the organization and operation of R&D institutions in the university. R&D organizations in production are not closely associated with the operation of the business. There is a large number of research institutions not associated with production and training, and businesses have no need for innovation. More importantly, Vietnam does not have a perfect market economy. This is the most important issue that leads to the nonexistence of competitive factors that force businesses to innovate.

The results of the study provide policy orientations for the development of Vietnam's STI system. The restructuring of the STI system is an inevitable objective on the path of reform from a command economy to a market economy. The reform needs to create an organic relationship between training, research, and production; establish the autonomy of the STI system; and restore the structure of the macro management system. The restructuring process should be suitable for the trend of international integration and especially for the early participation in the Bologna Declaration 1999. In the process of restructuring, the economy must soon move to a perfect market economy.

The book has outlined the theoretical and practical premises for STI activities in Vietnam, in the context of international integration in S&T. The results of this study describe the current state of the STI system, and these are placed in the context of international integration in order to show the gaps. The weak points of each element of the system are the macro management institutions. Recommendations for policy solutions and reconstruction procedures are developed to facilitate the compatibility of STI in Vietnam in the world's STI system in order to realize the integration process of Vietnam.

The proposed new S&T policy solutions only stop at the initial recommendations. Therefore, further research on this topic is needed to make practical contributions to promote Vietnam's international integration in S&T.

INDEX[1]

[1] Note: Page numbers followed by 'n' refer to notes.

© The Author(s) 2019
D. T. Truong, *Perspectives on Vietnam's Science, Technology, and
Innovation Policies*,
https://doi.org/10.1007/978-981-15-0571-3

Printed in the United States
by Baker & Taylor Publisher Services